Physical Intelligence

Physical Intelligence

The Science of How the Body and the Mind Guide Each Other Through Life

SCOTT GRAFTON

Pantheon Books
New York

 Published in the United States by Pantheon Books, a division of Penguin Random House LLC, New York, and distributed in Canada by Penguin Random House Canada Limited, Toronto.

Pantheon Books and colophon are registered trademarks of Penguin Random House LLC.

Library of Congress Cataloging-in-Publication Data

Name: Grafton, Scott T., author.
Title: Physical intelligence : the science of how the body and the mind guide each other through life / Scott Grafton.
Description: First Edition. New York : Pantheon Books, 2020. Includes bibliographical references and index.
Identifiers: LCCN 2019018374. ISBN 9781524747329 (hardcover : alk. paper). ISBN 9781524747312 (ebook).
Subjects: LCSH: Mind and body. Intellect. Thought and thinking.
Classification: LCC BF151 .G73 2020 | DDC 153—dc23 | LC record available at lccn.loc.gov/2019018374

www.pantheonbooks.com

Jacket image by edward way/500px/Getty Images
Jacket design by Linda Huang

Printed in the United States of America

First Edition

9 8 7 6 5 4 3 2 1

To Kim. For coming along, even when the shortcuts weren't.

Contents

Introduction

Danger keeps you on your toes.

—JIM BRIDWELL[1]

HOW DO YOU decide if you can drive through a snowstorm? How high are you willing to climb up a ladder to change a lightbulb? Can you prepare a dinner party for eight? When was the last time you discovered a shortcut through a forest?

For all these challenges, there is only one way to find out. A person needs to devote some time, energy, and physical engagement. Smart talk, texting, virtual goggles, reading, and rationalizing won't get the job done. The hands have to be on the wheel of the car to learn the feel of slipping tires. The feet need to be balanced on the ladder rungs to detect the tipsiness. The cook has to already know how to chop, fry, and combine four complicated recipes so they are all finished by a certain time. Best of all, finding a shortcut through the forest demands vigilance, courage, and the ability to keep one's wits, particularly at that moment of self-doubt when the journey seems more like a longcut than a shortcut.

Skills such as these are informed by "physical intelligence": the components of the mind that allow anyone to engage with and change the world. Inside the brain there is no single module or bit of tissue that makes this possible. Instead, the action-prone mind draws on a multiplicity of capabilities. This book is about these amazing mental operations, how they were discovered, and how they continue to be studied today. Some are almost primordial in their simplicity. How come you don't walk into walls or off of cliff edges? Others are quite subtle. When you take on a new do-it-yourself project, how much of your problem-solving relies on old habits, winging it, or careful reasoning?

Our psychological intuition about how the brain works inevitably places verbal thought and all the stuff we can talk about, such as our emotions, at the top of the heap. Physical intelligence, which is largely inaccessible to conscious introspection, is treated as a lower form of intelligence, something to be tucked beneath the verbal and largely ignored. This book makes the case that physical intelligence is much more. It is foundational, a kind of knowing that frames much of what the mind spends its time engaged in.* Indeed, the very fact that so much of physical intelligence can be performed beyond consciousness is the very design feature that frees a person's thoughts so he can spend his day thinking about social affairs, work, and the world of ideas. Under all the verbal chatter of the mind, much if not most of what the brain is actually dealing with is the raw physicality of being alive.

For many of my colleagues who study the mind, the very notion that physical action also requires some intelligence draws a blank stare. They focus on thinking and perceiving. Other than ears and eyeballs, the body is largely irrelevant for their kind of science. However, to study a mind without a body ignores some of the greatest pleasures of being alive: experiencing the world directly, as we perform and create. My

*It should be no surprise, then, that many of the greatest thinkers and writers get their best ideas while taking long walks in nature.

patients point this out to me time and again. As they lose various physical capacities they also lose bits of their deepest sense of self. One of my patients was a farmer in south Georgia with advancing Parkinson's disease. There came a sad day when I had to take his driver's license away. Driving has a way of projecting a person into the physical world, providing a dizzying sense of freedom. For good reason, then, the farmer was severely depressed when he lost his privilege. However, he was not to be deterred. Denied one of his greatest joys, he found an intimidating but satisfactory substitute: he could still drive his oversized bulldozer around his farm. For him, thinking, philosophizing, and reasoning would never offset the sheer joy of getting out and about in his vehicle. Even Stephen Hawking yearned for action. He once commented, "Obviously, because of my disability, I need assistance. But I have always tried to overcome the limitations of my condition and lead as full a life as possible. I have traveled the world, from the Antarctic to zero gravity."[2]

The hidden nature of physical intelligence poses a problem for the scientist. How can these capacities be exposed for what they are? To a certain degree, all of us are constantly searching for them. We are drawn like moths to a flame whenever we witness physical brilliance, when brain, mind, and body operate together with singular grace, as is sometimes evident in sports, dance, craft, or music. However, a scientist focusing only on superb physical talent can be led astray. It would be as if she were trying to understand language by only studying winners of spelling bees. All of our physical intelligence, not just that of outliers, needs to be explained. Look closely at the barista, the kid playing hopscotch, or the floor mopper and you will soon begin to notice physical brilliance everywhere. To show how and why this brilliance exists, I will dwell on some of my favorite experiments across a wide range of scientific disciplines, including the study of normal infant development, intracranial neurophysiology, robotics, brain scanning, and clinical neurology.

Long ago I discovered that some of the most important components of physical intelligence, the ones that are generalizable and relevant for all of us, are laid bare when one is alone in the natural world, particularly in the wilderness. Venturing into wild places requires enormous ingenuity and resolve. It is the primordial world we originated from as a species, and thus it makes sense that the cognitive capacities that are of greatest value for goal-oriented behavior should come to the forefront there. I make a yearly trip into the wilderness alone. I go to the Sierra Nevada, but one could imagine a similar trip in Alaska, the Rockies, the Cascades, the Okefenokee Swamp, or the great deciduous forests of Appalachia. This book is motivated by one of my trips and some of the capacities of physical intelligence that determined my fate along the way.

A good wilderness trip needs three things for the properties of physical intelligence to be evident. The first is obscurity. Although I had left a map and a detailed itinerary with my wife, I changed my route on the second day of my trip. If anyone went looking for me using the map I had given her, she would probably scour an area that was more than twenty miles away, beyond two glacial divides. Mobile phones don't work in these parts. And the Park Service is so understaffed, the likelihood of being rescued in a crisis is abysmally low. Without any of these lifelines, a relatively simple hiking trip can suddenly become a profoundly intense and complex experience. The second feature is solitude. On such a trip, there is none of the wonderful chatter and distraction that dominates the closeness and pleasure of an outing with family and friends. Without these entertaining social connections, a solo trip results in an utterly different kind of experience. It is not a lonely one. Rather, the solitude provides time for reflection and an opportunity to examine the kind of intelligence that informed human action as our species evolved. In addition, a trip alone completely changes the stakes and perceived risk. There is no confusion about responsibility. The traveler owns all his or her decisions. Roughness is the third feature. The familiar world is stripped bare; the setting is primordial. The landscape is open and stretches forever, with barely a trace of human influence. For more than 1.3 million years of evolutionary history, this was the ordinary world. There were no level sidewalks, warm houses, or high-rise luxuries. Nothing mitigated risk, eliminated hazards, or minimized effort. Our ancestors evolved in a world that was nothing but wilderness. This landscape endowed our species with remarkable ways of seeing, interpreting, and acting in challenging environs. With that in mind, when I take what are relatively hazardous and unknown explorations into the mountains, I get to experience a very crude simulation of what being alive was like long ago. Survival is paramount and one is ever mindful of it.

We didn't emerge as a species sitting around. We wandered far and wide, into locales that are almost unimaginable. To really understand physical intelligence, you need to wander. On a previous trip, I climbed one of the southernmost fourteen-thousand-foot peaks of the range I have spent my life roaming in. It was a very long, steep, ten-mile ascent requiring a wind-swept traverse far above the shelter of trees and running water. Just before the summit, I was surprised to find an obsidian arrowhead. The setting was desolate, remote, and cold. For many scientists, the arrowhead itself would have been critical. The object reveals something about the cultural intelligence of the hunter, his best technology, available resources, and trading network. The object could have been left there two hundred or nine thousand years ago. What interests me is not the specific manufacturing advances revealed by the arrowhead or how big the trading network must have been for it to show up in the region. Rather, it's the ridiculous location where I found it. The owner probably would have been a Paiute Indian who stalked a deer or mountain goat all the way to the summit, the far side of which ended in cliffs, effectively forming a trap. The hunter had readily climbed through this massive, unforgiving landscape at will, with stunning competence. To me he is amazing not for his technology (a stone-tipped arrow) but for his sustained confidence in stalking, tracking, and climbing over increasingly rough terrain while intensely exposed to sun, wind, or snow. All for dinner. When I roam through the middle of nowhere, the kind of intelligence he and countless generations of ancestors drew on becomes easier to appreciate. The point of my trips, then, is to wander through an environment that makes the natural relationships between thinking and acting obvious to anyone.

To get a good glimpse of what people were like when physical intelligence was honed, one has only to look to the "Iceman," Ötzi, a mummified hunter who died five thousand years ago just below a high mountain pass on the current Austrian-Italian border. Ötzi's remains are on display at the Südtiroler Archäologiemuseum in Bolzano, Italy, along with his fur-covered bearskin hat, goatskin clothing, copper-tipped axe, backpack, food scraps, medicine, fire-starting tools, and longbow. Only five feet three inches tall and with an entirely ordinary albeit fit physique, he readily traveled alone and self-sufficiently through the middle of the Alps. He could smelt copper. Take down a large elk. Fabricate Neolithic blades. Travel over glaciers without getting frostbite. He yields a startling view of our past. What we consider to be extraordinary—living in an Alpine wilderness—was entirely ordinary at the time. Ötzi relied on a complex of mental capacities that allowed him to adapt his behavior constantly to meet the demands of an extreme and highly varied environment.*

The chapters that follow are a sampling of a few of these capacities—basic properties of the mind that not only kept ancestors like Ötzi alive but continue to manifest themselves in all of us. A common thread running through the chapters is the special role that learning plays in forming this kind of intelligence. The mental capacities that are used for action are, more than anything, different kinds of learning machines that the brain has available for acquiring and maintaining physically derived knowledge. Physical intelligence is absolutely ruthless in requiring that knowledge be gained from direct physical experience. This is profoundly different from, say, the instantaneous remembrance of a face, name, or phone number. Rather, physical intelligence reflects learning processes that constantly tinker with a person's performance. One never stops learning to cook, to drive, or even to walk, for that matter. It is also a knowledge that is lost from disuse: without practice you will fall on ice or off ladders.

The world Ötzi navigated was physically challenging and complex, characterized by palpable tension arising from an inability to predict what might happen and few means for maintaining control. Here were perfect conditions for improvising, inventing, and enduring some of the most rigorous demands of the wilderness, which lay at the heart of what shaped physical intelligence for eons. Although the wild is uncontrolled, physical intelligence provides the means to establish a sense of control. Humans acquire their skills and learn to solve problems through constant physical experimentation. That was as true

*Ironically, what led to his demise was not the rigor of an extreme mountain environment; it was someone shooting an arrow into his shoulder.

for Ötzi as it is for us. There is no end to the sensing, adapting, anticipating, and accommodating that must take place for a person to act intelligently. It takes practice and know-how to do even the little things in life: to stay upright on a slippery sidewalk in front of your home or to know whether you can still climb a ladder without falling off. And most of all, physical intelligence provides the means for experiencing the pure joy of figuring out how to do something for the first time, whether it is building your first campfire or catching your first fish.

Physical Intelligence

1

The Space We Create

The more space and emptiness you can create in yourself, then you can let the rest of the world come in and fill you up.

—JEFF BRIDGES[1]

FROM THE DRY scrub desert in eastern California there are only half a dozen roads winding into the High Sierra. They rapidly gain a mile or more of vertical elevation before abruptly ending in tight slot canyons, narrow V-shaped wedges cut out of the surrounding peaks, crowded by vertical bands of granite cliffs and endless piles of rock. The walls of each canyon bulge in on three sides, blocking each afternoon's rapidly setting sun. Each has a full creek ripping down the center, with willows, pine, and birch trees hugging the banks and the surrounding underbrush blocking any view through or out of the canyon to the flat valleys far below.

After driving many hours in the open scrub desert, I found this confined terminus claustrophobic. I unpacked the car in a pleasant little U.S. Forest Service campground supporting only five tent sites tucked in among a circle of Jeffrey pines. Each site has a parking spot, a picnic table, a firepit, a bear box, and a flat patch of leveled-out gravel to pitch a tent. There was only one other group around that night, four jovial guys enjoying a twelve-pack while stoking an unruly campfire. I pitched my tent in the farthest site from them so their voices would be muffled by the pines. I took everything remotely edible out of the car and locked it all up in the bear box. Besides food, this included toothpaste, medications, zinc oxide, sunscreen, camera (with a delicious leather case), and, of course, food and cookware. It was a drought year, and even worse, summer was quickly coming to a close. Thus it was a last chance for bears to stock up on calories before their winter slumber.

I settled into my sleeping bag and stared toward the ceiling of the tent. A storm had blown through over the previous week, and at 9,200 feet above sea level the cold came in fast after sunset. There was no point standing around outside shivering under a moonless sky that was quickly turning pitch-black. I began to unwind after six hours of driving. I could feel myself making the mental transition that happens every time I head into the mountains alone. My sense of awareness, the space I attend to in the environment, was steadily expanding, stretching far beyond the flimsy nylon tent fabric, stretching to encompass much of the canyon grotto. Rather than mentally jumping from sound to sound, as a person might do trying to catch conversations at a cocktail party, I was listening to nothing in particular, taking it all in without focusing on any one sound over another. As it turned completely dark, sounds became more alive, more salient. I wasn't just listening, I was vigilant, reaching out into the space all around, trying to get as much from it into my awareness as possible.

The vigilance I deployed is at the heart of any animal's physical intelligence: a detailed sampling of the environment to stay ahead of danger, to predict and to respond. Vigilance like this, the kind that you rely on to stay alive, is quite difficult to study in a laboratory. What gets studied instead is a boring kind of vigilance: a person's capacity to stay awake while performing an exceedingly tedious chore. For example, a person might be asked to stare at a blank computer screen for hours on end, tasked with pushing a button whenever a plus sign unexpectedly appears. This is the dreariest of sentry duties and, not surprisingly, many experiments show that when a person is asked to do a pointless and boring task, drowsiness and performance failures will ensue. Contrast this with the vigilance I was employing. There was nothing boring about what I was doing that night. I was excited to be in that wild canyon to experience whatever might happen.

There is another kind of vigilance that gets tested on animals, mostly mice and rats. In a typical experiment, the animal is put in a cage where it has already learned it will get an electric shock if something specific occurs, like an acoustic beep or a flash of light. Not surprisingly, its behavior changes as it anticipates the upcoming electric whack. While this kind of experiment does a better job at bringing danger into the equation, it still falls short of characterizing natural vigilance because the entrapped animal already knows the guaranteed injury to anticipate. The tortured are not vigilant. They are terrified. In contrast, in my tent, listening to what might be present all around me, I had only a vague notion that there might be danger, albeit at a very low level of likelihood, and I had many options for doing something about it. I was in a state closer to what you might find yourself in while alone in a sketchy parking lot late at night. While there is no clear and present danger, you might get the sense that there is something wrong. Your awareness expands as you simultaneously monitor for footsteps or gruff voices in a distant stairwell or behind some parked cars. Vigilance in the parking lot is all about understanding the large surrounding space, being in the moment, and imagining possible threats. Vigilance like this, whether in a parking lot or in the wild, is an automatic and effortless cognitive competency.

The vigilant mind requires something very fundamental of the brain: it has to create a concept of space to be able to both understand and act in the physical world. Just as soldiers, firemen, and search-and-rescue teams work within an operational space that they draw on a map, the human brain creates a three-dimensional space to compute, sense, and act within. As I listened, the operational space that I mentally tracked extended far beyond the confines of what I could see inside the tent, to the farthest points of the campground. Our ideas, images, and actions are all organized within some sort of operational space, whether it is the inches defined by a looming kiss or an area the size of a football field. There must be a "where" for vigilance to work. Creating and manipulating mental space is done gracefully and without any conscious thought. Thus it is easy to take something as abstract as space entirely for granted and to ignore it entirely. And it is really hard to study in the lab because it is, by definition, a nothing within which something might happen.

The brain routinely combines vision, touch, and positions of the joints to make a volume around the body. There is good evidence that some neurons code for particular parts of space, such as the space within reach.[2] Other kinds of neurons are active when an object or the hand approaches a particular spot on the face, like one cheek or the other.[3] The activity of the neuron is like the alarm on an automobile's proximity sensor, blaring out a warning when the driver gets too close to another's bumper. These neurons are creating a volume of space for actions up close and personal. We know far less about how the brain creates an understanding of the rest of what is out beyond our reach. Nevertheless, we know the brain must make all kinds of space, and when it can't, the deficits are profound.

Imagine visiting an elderly relative at the hospital. She has a clinical story that is typical for patients on the stroke ward. Two days earlier she had stopped answering her phone, and so you called the paramedics. When they checked in, they found her unresponsive on the kitchen floor. After arriving at the hospital, the neurology team discovered she had a stroke due to a blood clot in an artery in the right hemisphere of her brain. After a scary overnight vigil, you are delighted by the good news that she woke up and is talking and moving her arms and legs. You rush to her bedside, hold her left hand, and tell her how glad you are that she is doing better. But something is amiss. She completely ignores you and continues staring at a television. No matter what you say, she is blandly indifferent, as if you didn't exist. Frustrated, you go to look out the window on the other side of the bed. "Oh, hello!" she blurts out. "I was hoping you would come to see me." Confused, you nevertheless enter into a pleasant conversation to learn how she is doing. She complains about feeling tired and that she wants to go home. When you walk back to the other side of the bed, to her left, she ignores you again. What is going on here? To her right, you are her favorite friend, and to her left, you have vanished. Don't take it personally. Neurologists have labeled this syndrome "spatial neglect"; it is as if she is ignoring part of the space that surrounds her.[4] However, for severe deficits in patients like this, the terminology is misleading, even unfair. Ignoring requires intentionality, but there is nothing intentional about what she is doing. The patient you are visiting has a problem that is far outside of her control or intention.

What is startling about patients with severe neglect like hers is that her loss of space can be so absolute. Look carefully and you will find she also fails to make any goal-directed actions into that neglected space, whether it is turning her body, gazing, or reaching toward that side. It is a complete void with no content, as if that part of the world was pulled into a black hole and is no longer reachable by any means.* Of course, the real world is fine. Her brain just can't construct a placeholder for the missing part of space; it is something that she simply cannot create in her mind. And if the brain can't create some space, then nothing else follows. Neither your words nor visual presence can escape her black hole of space and reach the rest of her mind, where there is some space to work with. She can't really be accused of intentionally neglecting or ignoring you when you are located in a part of space that from her perspective doesn't even exist. She reveals how profoundly elemental the construction of space is, from which all action in and interaction with the world follows.

We intuit that space is constructed like a very homogeneous three-dimensional volume that envelops us and is filled up with stuff. However, other stroke patients with variations of the neglect syndrome reveal something more nuanced: there are a bunch of different ways to construct space. For some neglect patients, it is not just the space around them that might disappear, it is the space containing their own body that disappears. They can't be expected to know they have a paralyzed arm when that part of their body no longer exists in their mind. Or they may lose bits of space that envelop objects in the world. While on clinical rounds I would test this by asking everyone on my team to wear his or her white coat over the left shoulder with the rest of the coat behind. A patient with what is called

*There are patients with far less severe neglect who may not completely lose spatial awareness. Rather, they will be able to perceive and act in the neglected space but not as well as they do in the intact space. Regardless, they have a poorly formed operational space that undermines any expression of physical intelligence.

"object neglect" would be unaware that we were wearing our coats only on one side. The space of each object—that is, each doctor or student—wasn't being created to the patient's left, and so the coatless side of each person fell into a separate black hole, ceasing to exist in the patient's mind. It is startling that so many small instances of space, each encompassing a different object, could simultaneously vanish. Other patients, when seeing a place in the mind's eye, have difficulty imagining what lies to the left side of a scene. Here we have a black hole in the space of ideas within the mind. Considering these different neglect syndromes, one understands how fluid the construction of space really is.

The insights into how the brain makes space come mostly from brain scans of patients. The scans localize where neglect patients have been injured by stroke, consistently identifying very distinct brain areas that enable the creation of space. Typically, neglect arises when there is damage to a zone at the junction of three right-brain areas: the parietal, temporal, and occipital lobes along with the underlying wiring that connects them to one another. Big lesions eliminate space altogether. Smaller strokes to these brain areas can distort space in amazing ways. Patients can lose depth perception, be unable to recognize 3-D structure in a scene, and even mix up the right and left sides of a scene.[5] Space itself can become skewed, leading some patients to tilt to one side in the aptly named "Pisa syndrome"[6] and others to lean their body toward the distortion in the "pusher syndrome."[7]

The confined localization within the brain for making an operational space suggests that this capacity should probably be thought of as an elemental brain process, a backbone for all sorts of brain tasks like vision, audition, or movement. In terms of comparative anatomy, the cortical zone that creates this sense of space is vastly expanded in human brains compared to those of other nonhuman primates. Assuming bigger is better, the case could be made that the ability to create space in our heads, while not uniquely human, is nevertheless particularly robust, providing us with a large and varied operational theater to work within. From a computational perspective, the way that neurons actually accomplish this amazing cognitive trick remains a mystery. Artificial intelligence can't come close to matching the human ability to conjure a flexible virtual operational space.

Although it remains difficult to determine how the human brain actually creates mental space, one thing cognitive neuroscience has nailed down is the way that a brain can use attention to control what will fill up space and to move the mind through space. These two ways of using attention, the what and the where, were central to how I treated space as I lay in my tent. As the sun set, I shifted my attention from looking around at the walls of the inside of my tent to listening to the rustling sounds in the bushes outside. As I moved from inside to outside I shifted from sight to sound. This was a smooth, effortless transition from what to where in space, a capacity that is a core property of physical intelligence. There is magic in this ability to control whatever bits of space we want to attend to: they can be shrunk, stretched, or shoved to the side. Measuring how attention manipulates space often starts with measuring how well a person can stay focused on one particular thing or location in space, the familiar admonition to "pay attention!" This ability is captured in experiments where the participant stares at a computer screen and searches for something in particular: a stripped-down version of Where's Waldo? The metaphor for what the brain is doing in this case is the zoom lens. Its job is to search through space to find something specific and, once found, to focus on the target. It does this by searching for features (like Waldo's face) and by examining only a little space at a time. Within this spot, information is amplified. These kinds of experiments consistently yield two insights. People can readily zoom their attention into small areas of space, and they can do this without physically directing their eyes or head toward the location. In my tent, I could track the location of a mosquito by moving attention around while keeping my eyes fixed on a spot on the ceiling.

The flip side is a form of attention needed to filter out unwanted information. A person strolling down Fifth Avenue in Manhattan is bombarded by both the press of pedestrians and bright, shining objects in store windows. Imagine if you were unable to ignore all the window displays on Fifth Avenue. You wouldn't be much better than an excited puppy, running from item to item. And yet, a person talking on her cell phone to a lover is able to pass by all of it, oblivious to the visual clutter. The capacity to filter selectively different sensations that are being dutifully recorded by our eyes and ears is an extraordinary evolutionary accomplishment. At the same time, there needs to be sufficient leakiness in this filtering operation so that conspicuous events or noteworthy objects can be noticed. Your passionate phone conversation will almost certainly be interrupted if there is a car crash in front of you. There is a delicate dance between filtering out what is irrelevant and being ever sensitive to salient events in the world.

This also starts to get us closer to a realistic notion of vigilance. One type of experiment that tests this operation in the lab challenges people to search for information on a computer screen while distractions are flashed on the periphery. A good analogy is the chore of using a nasty Web browser loaded with pop-up ads appearing in the margins. For the researcher running an experiment, as well as the ad agency, the name of the game is understanding what kinds of features are most effective at breaking down a person's ability to filter out irrelevant information by manipulating the salience of the distractions. Which pop-up ads are most effective? Usually the most irritating, the ones that unexpectedly appeal at once to the eye and ear or that are particularly loud or bright. One can readily find parallels in nature that drove evolution to treat some features as more salient than others. A rustling of bushes, a slithering motion on the trail, a howl in the night. Whether in the lab or in the wild, the brain must not only filter out irrelevant stuff but at the same time decide when to shift attention to new information.

Filtering the perceived world requires some sort of decision about what is salient. Salience is defined as the quality of something being particularly noticeable or important, and it is the second part of this definition, the arbitrariness of something being important, that makes "salience" so slippery and yet so essential a term. Somehow, the perceiver has to know ahead of time what might be important. And yet, brains are superb at knowing what should be salient in most situations. In my tent, the sound of a nearby cracking branch has incredible salience. Is it a bear? But if I was in the same tent located in a crowded Yosemite campground with people and raccoons constantly trudging past, then the crackling branch might have no salience at all. This flexibility for redefining what is salient within an operational space is a remarkable achievement of neural coding. It requires more than just a knowledge of what is likely to be present or absent in the environment; it also considers what the person has as a goal. Salience mediates how a person detects and attends to all sorts of goal-relevant inputs and adjudicates what to ignore.[8] It also comes into play whenever there are goals with competing choices. If you are out shopping and trying to decide between two fine pairs of shoes, it prevents your attention from being entirely captured by one pair over the other. It is easy to see how salience plays a central role in how we apply our thoughts to the world. But what in the brain actually controls salience and attention? Is there an executive function that controls attention in all of its complexity?

In the early days of cognitive neuroscience, the mind was viewed as being built from a set of modules, each located in a specific part of the cortex.[9] This view grew from foundational studies of stroke patients, dating to the 1870s, many of whom had strikingly specific deficits and circumscribed lesions.[10] This modular view was reinforced by many functional brain-scan studies, which mapped specific tasks like seeing, feeling, moving, and remembering to distinct sites in the cortex. From this, it was tempting to believe that there would be distinct modules in the brain for controlling attention from the top down, including a brain area for controlling salience. The conceptual problem with this is in nailing down what is actually meant by "top." What part of the mind, if any, will always be at the top to maintain top-down control? It is easy to imagine a model of the brain organized like a corporation, with a CEO module at the top, delegating a staff of middle-manager modules to do her bidding, the managers maintaining attention or running filters. However, try as we might, our brain scans never could find the homologue of the brain's CEO module. Rather, beginning in the early 1980s, a profoundly new view for understanding the delicate balance among attention, salience, and space began to emerge. In a highly influential paper, the neuroanatomist Marsel Mesulam made a drawing, a sketch really, showing how areas that regulate attention were spread far and wide.[11] There was no single hub for controlling attention. Mesulam's drawing, decades ahead of its time, eventually evolved into a network diagram that has become a standard model for how attention might be regulated.* Since then, thousands of functional brain-scanning experiments have clarified that visual attention, which can be directed willy-nilly, is regulated by a syndicate of cortical areas spread across the upper regions of the brain (that is, in the dorsal direction); these areas are collectively referred to as the "dorsal attention network." The network is tightly linked with brain areas that control where the eyes will be directed as well as areas of the frontal lobe that are essential for maintaining mental control of one thought or action over another. An entirely separate syndicate of brain areas located along the lower (ventral) aspect of the brain influences what gets filtered out and what gets through. Because of its location, it is called the "ventral attention network." Ultimately, a third syndicate of brain areas called the "salience network" was defined. It manages the flexible allocation of attention between the dorsal and ventral attention networks and integrates attention with goals. It does the balancing act of deciding whether to focus, whether to filter, and what to filter for. Consistent with Mesulam's proposal, there are dense anatomic connections among these three networks. There is no "top" module that directs a person's attention. Rather, the contribution of these networks is best understood by mapping their functional interactions and determining at any given moment which of them is most influential. This is done routinely using a brain mapping method called "resting-state

* Mesulam's paper is a landmark in using, for the first time, the term "network" to describe the large-scale anatomic organization of a specific functional system in the human brain. He deduced the network's structure by directly observing the patterns of connectivity in the deep white matter of the brain, the areas they projected to, and the locations of strokes in patients. It would take two decades after his publication for the concept of a network to become widely adopted as a distinct unit of anatomic description.

functional imaging." In these experiments, people are asked to lie in an MRI scanner and their brain activity is measured while they do nothing in particular. They spontaneously think, plan, daydream, or recollect recent events. The degree to which activity is coupled between areas is a good indicator that they are functioning together. One finds strong interactions between the salience network and the other networks. The salience network also stands out because its connectivity to the two attention networks fluctuates rapidly, as if capturing fleeting shifts of attention and thought as a person jumps across a landscape of ideas.[12]

The notion that the regulation of attention and salience is widely distributed in the brain is borne out in further studies with stroke patients. While there is a place in the brain where damage will destroy the ability to create space, there is no analogous spot in the brain where local damage will eradicate the control of attention or the ability to integrate salient information into the shaping of attention. Instead, the typical changes of attention in the neurology clinic after a brain lesion are relatively subtle and can emerge with injuries located in many different places. You may know someone who sustained a concussion from a car crash or sports accident. His clinical MRI scan may look perfectly normal. He can still regulate his attention, but not with the flexibility and sensitivity that one would normally expect. He has problems staying focused on one task or another. He may not like distraction. Or he may get quite upset if you ask him to drop one task to start another. He benefits from an imposed structure or schedule. Functional MRI scanning (fMRI), the type of scan we do in the research setting, reveals a possible explanation: he lost some of the functional connectivity across the attention and salience networks, a little bit here or there, leading to a graded decline in network dynamics and cognition.

Laboratory experiments with attention reveal the superb ability of the brain to zoom in on, amplify, and filter information within almost any sort of space. As I lay there in my tent, I couldn't help but think that these capacities capture only part of the story about how the brain manipulates space. I was listening to as many sounds as possible in as large a space as I could project outside my tent, a space through the trees and beyond the campground. Although I was listening, not looking, I was manipulating space as if adjusting a camera lens beyond the wide-angle setting all the way to the fish-eye setting. I could simultaneously track a nearby brook, a remote torrent, and the laughing voices of distant campers, no one sound grabbing more attention than any other. What I was really doing, of course, was implicitly monitoring for anomalies, salient unexpected sounds that didn't fit in.

For all the thousands of experiments that have tested how visual attention can be manipulated like a zoom lens, the number that consider whether attention can also stretch space like a wide-angle lens can probably be counted on two hands. The fundamental question of course is whether people have an ability to intentionally widen attention. Experiments concerned with visual attention show how this might be possible. If a person stares at a dot on a screen and tries to detect targets in her peripheral vision, her performance gets better if she intentionally widens her focus of attention,[13] while performance outside of that area declines.[14] But do these adjustments require the brain to push attention outward or does the brain naturally assume the widest possible field of view? Unlike a zoom lens, which works automatically when there is a particular object to point at and zoom in on,[15] the converse is not necessarily true for a brain's wide-angle lens. There is emerging evidence that the background in a scene is particularly impactful at setting the amount of wide-angle attention. In other words, adding interesting things to the background is an effective way to trick attention into encompassing a wider field of view, without the person even realizing it.[16] Another way to think about wide-angle attention is to ask if there is such a thing as background consciousness.[17] Experiments show that just as people can consciously center and zoom visual attention onto objects or specific locations, they can also be highly attuned to the visual background, and this in fact can be a default mode for analyzing a particular scene. This background consciousness has been proposed as an early warning system for anomalies in the environment. My experience in the tent suggests that in a given context, like the possible presence of a bear, background consciousness can readily be brought to the forefront of awareness.

Researchers are still working to find a brain network that supports wide-angle attention or sustains background consciousness, but most likely it is the ventral attention network. When a person in a functional brain-scan experiment simultaneously has to pay attention to events that could occur either to the right or left there is greater activity in the ventral attention network.[18] Other experiments using pictures of natural scenes show that the ventral network is involved in on-the-fly decisions about when to cease attention on one object or location in favor of another, a process that is central for maintaining vigilance.[19] The ventral attention network is also linked to awareness, no matter where a person is focusing attention.[20] These different experiments show how the ventral attention network likely plays a central role in the expansion of space, holding on to information within that space and having awareness of salient events or patterns there.

It is noteworthy that macaque monkeys, sophisticated aerial masters of the forest canopy, have only a rudimentary dorsal network compared to humans. In contrast, they have a robust cortical system corresponding anatomically to the human ventral attention network.[21] Their typical behavior in the wild could be interpreted as a reflection of this relative anatomic imbalance: they have far less volitional control over where and what to attend to compared to adult humans. At the same time, they are highly responsive to potential threats in their environment. It seems likely that many primate species developed adaptations providing for an enhanced ability to monitor the background, perhaps with the ventral attention network serving as an early warning system. Only later was there an emergence of the capacity to control where attention could be pointed independent of all the potentially salient stuff in the environment.

Recent studies are making the case that a similar imbalance between attentional systems also occurs as part of normal adolescent development and may explain some of the symptoms in attention deficit disorders. As many frustrated parents will tell you, adolescent boys are short on their capacity to focus and sustain attention on what, to them, seem like behaviorally irrelevant tasks. In this light, their distractibility is remarkably similar to that of the macaque monkey. The tough-love reaction of many teachers and parents assumes the students are too busy mind-wandering and just need to buckle down. Instead, it is more likely that they are attracted to distractions in the background because that is the only capacity that has fully developed in their brain. For all practical purposes, they are like a monkey in the treetops, readily distracted by the background due to heightened monitoring.[22] To test this, resting-state brain activity was measured in children aged seven to twelve years.[23] Compared to adults, the children showed greater functional connectivity between portions of the ventral attention network and nodes of the salience network, consistent with the neural signature from monkey imaging studies and supporting the idea that this ventral network and its sensitivity to salience or a heightened state of vigilance are dominant in youth, with the dorsal attention network needed for sustained attention appearing later. Thus there is a mounting case that neither monkey nor boy has the neural circuitry for sustained, focused attention. For the monkey, it may never be particularly robust. For the boy, it takes a few more years of development of the dorsal attention network. This brings up the question of whether some children labeled as having attention deficit disorder (ADD) have a real brain disorder or are simply misplaced with respect to their stage of cognitive development. For those children with ADD dominated by hypo-attentiveness (as opposed to hyper-impulsiveness), there is relatively increased connectivity throughout the ventral attention network.[24] Furthermore, there are weaker connections among the salience network and cognitive control and the dorsal attention network.[25] In this light, the inattentive boy is not abnormal, in need of Ritalin or a psychologist. He is a well-adapted young member of his species, not yet sufficiently muscular or skillful to be independent. Situate him in a prehistoric setting and it would be critical for him to be vigilant for hazards in his environment.

When I go into the mountains alone, the dynamics of the environment and potential hazards are, respectively, relatively slow and easy to predict. Nevertheless, I need to be vigilant. I don't know what will happen around me, but I have to find some way to predict what might. Rocks move, trees fall, snow avalanches, ice breaks, winds gust, snow bridges collapse, horses kick, bears move in, rattlesnakes shake. The stakes are high. If I miss something, I can end up in a real pickle. Of course, if you live in the mountains, this ability to monitor the environment and predict what might happen is learned naturally through lifelong experience. However, most of us enter into risky environments like this only rarely, swiftly transitioning from safety to the unknown. How do we train our attention and state awareness so that they are sensitive to threat or danger, particularly when we spend most of our time in our predictable suburban or urban environments? In the modernized world, team sports afford a way. The ability to expand attention to a larger operational game space can dramatically augment performance. Wayne Gretzky, one of the greatest hockey players of all time, was known for his astounding on-ice situational awareness. One might think that he was very good at quickly zooming attention to different points on the rink. However, in his own words, whether he was skating with the puck or not, over a lifetime of practice Gretzky taught himself to not focus on specific features in the games, like the uniforms of teammates or opponents. Instead he relied on quick glances to pick up colors and designs, just enough to update an ongoing mental map of the game. Gretzky has been quoted by sports pundits as saying, "If you ask a 50-goal scorer what the goalie looks like, he'll say the goalie's just a blur. But if you ask a five-goal scorer, he'll say the goalie looks like a huge glob of pads. A five-goal scorer can tell you the brand name of the pad of every goalie in the league. I'm seeing the net, he's seeing the pad."[26] Gretzky thought his ability came about as a survival skill. "When you're 170 pounds playing with 210-pound guys, you learn to find out where everyone is on the ice at all times."[27] When star athletes are constantly shadowed and checked, they are forced to keep their eyes open and head up, and they naturally develop an acute awareness of their surroundings. They are forced to monitor the periphery—*i.e.,* the background—far more than their competitors.

There is a long-standing debate about whether the capacity for situational awareness in athletes like Wayne Gretzky is learned or a genetic gift. The most compelling case that vigilance and a widened sense of awareness can be learned is readily demonstrated by combat-experienced soldiers. A few years ago, I ran a brain imaging experiment that sixty-five U.S. Army officers participated in. All of them had been on multiple tours of duty and had seen extensive combat. They were back in the States training new recruits in their lessons learned. Normally, work at the UCSB Brain Imaging Center involves a lot of undergraduates and nerdy graduate students. Things were dramatically different around the center when very large, intense army officers started showing up. What everyone on the project noticed was the officers' habitual scanning of the environment. They would systematically gaze at locations around the courtyard outside the lab that an undergraduate wouldn't even notice. Their continuous, automatic vigilance is a clear adaptation of physical intelligence, inculcated in them both formally and through hard-fought experience. There is also an emerging theory that too much of a good thing can lead to dysfunction. Specifically, there is strong evidence that many (but not necessarily all) veterans with post-traumatic stress disorder have persistent and maladaptive vigilance as demonstrated by a difficulty in disengaging or releasing spatial attention.[28] In the same veterans, MRI measures of functional activity across the brain show that they have stronger connectivity between the salience network and the two attention networks, perhaps revealing an inflexibility in the way their brain regulates dynamic activity across these networks. This theory has led to innovative clinical trials that focus on training methods that modify a PTSD patient's control of attention. This could be done either by training attention in a general way or, alternatively, by desensitizing a veteran to threatening or nonthreatening stimuli.[29] Results so far suggest that the greatest benefits come from interventions that improve the patients' overall control of attention, rather than training that teaches them to ignore or attend to particular threats.[30]

For those who don't play team sports or wander in the

wilderness, there are other avenues for enhancing awareness and vigilance. The closest urban experience might be mindfulness training, whether this is accomplished through formal meditation or closely related techniques that seek to create the sense of being present through body-scan awareness exercises, sitting and walking meditations, mindful eating, and mindful movement such as yoga. At first blush, this choice of training might seem a bit paradoxical. Mindfulness exercises are performed in some of the least threatening environments ever created (temples, monasteries, salons, retreats). How could quiet meditation lead to keener awareness or vigilance in harsh, sometimes violent environments? There are several competing theories about why such training works. One idea is that this type of training can lead to improved ability to control attention, particularly focused attention, via the salience network. Many sects of Buddhism and styles of meditation are built on focusing the mind on a single thought or point in space. For example, a common Buddhist meditation method teaches people to consistently focus attention on one thing over sustained periods of time. This can be one's breath, a feeling, one's heartbeat, or even something external. Try this, and you quickly discover how easily attention wanders. The challenge is to detect the wandering and refocus. In theory, with enough practice, a meditator develops a relatively effortless ability to regulate attention, whether by detecting distractions, letting go of them, or getting back to the object of focus. For many other practitioners, there is a complementary form of meditation, called "open monitoring," where the aim is to be in a continuous attentive state, monitoring everything that occurs without focusing on anything in particular.[31] In this way, the use of an object as a primary focus is replaced by effortless sustained awareness without explicit selection. A core feature of open-monitoring meditation is "reflexive awareness," presumably a state that allows for a stronger sensitivity to natural phenomena, richer emotions, and more active thought. Mindfulness training is similar in providing the means for people to increase awareness of their surroundings, to be in the moment. The techniques of open-monitoring meditation and mindfulness are probably as close as one can get to the vigilant mental state that emerges in the wild. Unlike focused attention, where the environment is largely filtered out, here there is room for a person to take it all in, to experience her environment without prejudice.

A growing scientific literature is establishing fascinating similarities between modern meditation and mindfulness techniques and the mental states associated with heightened vigilance. The physiology of the brain in open-state meditators has been recorded by electroencephalograms. Their meditation is accompanied by very high-frequency rhythmic brain activity, particularly between parts of the salience network and the sensory cortical areas.[32] Experiments in mindfulness training have recruited people with high levels of psychological stress resulting from unemployment and other life challenges. In one study, participants underwent three full days of mindfulness training and were compared to a control group at the same retreat who were taught relaxation techniques.[33] Those with mindfulness training demonstrated far greater clinical benefit with reduced stress. Notably, a comparison of brain scans measured before and after mindfulness training showed a strengthening of functional connections between areas involved in cognitive control, salience, and the ventral attention networks, changes that did not emerge from relaxation techniques alone.

The great irony, of course, is that as a species we did not evolve to meditate or to be mindful. These are recent cultural inventions. There is simply no survival advantage to sitting and contemplating the universe. So, from an evolutionary perspective, what is the vacuum that people are filling with their mindfulness and meditation? I believe that it has become a proxy for a desired mental state that cannot be easily triggered in modern environments and that normally arises when people deal with natural unpredictability. In this light, the meditator is using extensive training to reboot intrinsic cognitive abilities but doing so in modern safe contexts. Go to the mountains and these abilities arise on their own.

It is well past dusk and I start to doze off. I awake to frantic shouting emanating from the other campsite. Flashlight beams are whipping around as four panicked campers holler and clang pots together. It is a familiar chorus, played out every year in campgrounds throughout the Sierra. It is a song to accompany a bear as it cleans out the singers' grub. It is clear from the desperate cries that the campers hadn't considered the possibility of bears and left an ice chest out. Their chorus went on for a long time, to no avail. Bears are meticulous eaters and can't be rushed. As I listened to the munching through the din, I was impressed at the bear's relaxed ability to stay focused on its task. Eventually the eating stopped, the flashlights went out, the banging stopped—no more bustling and noise. Now with the intrusion ended, the campers went to bed. Then it really got quiet, with just the background rush of the distant creek and a soft wind in the pines. Amid the calm, I considered whether I had locked up all my food. You never know. Might there be anything edible in my coat pocket? Were there old cracker crumbs in the tent? My attention widened into the opaque night. I was wide awake, monitoring my environment as with a fish-eye lens. Open, attentive to everything, a Zen moment achieved without mindfulness training.

2

Surfaces

Nature is a language; can't you read?

—MORRISSEY[1]

THE EASTERN VALLEYS of the Sierra are greeted by a rapidly rising summer sun that warms the trails to a baking temperature by midmorning. To beat the heat, I pushed myself out of bed before dawn, had a quick meal, and headed uphill to cooler elevations. The trail was popular for its relatively quick access to the inner mountain wilderness. Along the way it climbs a steepening canyon dotted with trout-filled lakes hugged by mossy meadows and massive pines providing sporadic shade. But speedy access has a price. The trail, created more than two hundred years ago, was designed for shuffling pack animals rather than humans. The path is cruelly monotonous, with little variation in step, steepness, or pace. The same muscles drive for each step. This is the kind of regularity that wears a person out, that makes him notice the weight of his pack and the discomfort of the tiniest pebble in a shoe. Between occasional vistas, the trail required that I keep an eye on the same dull view of the ground below. That is the trade-off of a well-groomed trail. It allows speedy travel, at the cost of enjoyment. One feels like a forlorn pack mule in this setting.

Fighting the drudgery, I tried to pass the time by playing

mental games with my altimeter, measuring how fast I was ascending, predicting when I would reach a pass at twelve thousand feet above sea level. In the mountains, increments in altitude, not mileage, are the real measure of progress. After an hour, I reached an elevation where big trees could no longer survive and transitioned into a series of lovely alpine meadows interspersed with gnarled white pines, mere dwarfs compared to the giant trees of the lowlands. With each new bit of elevation, the meadows thinned until I was mostly traversing through rubble. After another hour, I was delighted to reach the pass. A Sierra pass is always exciting because it opens a unique portal into a new and unknown place. One feels an anticipation, like stepping over the threshold of a grand opera house or a cathedral. There is surprise and wonder at the architecture on the other side of the door, instantly transforming mood and thought to elation. I gazed at a distant line of serrated mountains and a foreground of jumbled terrain. Crossing this pass also signified a release from the tyranny of the pack trail. From here on, my travels would be off trail, a pathless cross-country ramble on whatever surfaces might present themselves. I took off across a scrubby meadow, heading for a broad saddle that was five miles away. Once free of the trail I found I was a bit clumsy for the first mile or so, easily tripping or misstepping. It took time to adjust pace and balance in this rough-and-tumble landscape, to find my "sea legs." Unlike on the groomed pack trail down below, no two steps were alike; every muscle was being used. Vision had to be allocated more dynamically, quickly shifting between the ground in front where I would land my next step and the landscape far ahead, to plan a route over an increasingly ragged terrain. There were constant route choices as I weaved around crags, up bluffs, and over small streams. A slope might be too slippery, a cliff too tall, or bushes too thick. I soon had my rhythm back, delighted with the task at hand. The complexity of the terrain enforced absolute concentration at each moment; it was the essence of what pop psychologists have referred to as "flow state." Unlike the pack trail, there was nothing dull or monotonous. Off-trail rambling is primordial, a physical skill we have used for as long as we have been a species, and showing it off can still be immensely pleasing.

To move through the world like this as one desires, the mind has to recognize what is possible or impossible. It takes some effort to grasp how acutely the brain tracks the surfaces of the physical world, the literal hard constraints that limit what is possible. And there are loads of these surfaces. Imagine taking a walk through a majestic California redwood forest. What do you see? You might describe a few of the trees, the noise of the birds, unusual ferns, or a cloud in the sky. These verbal descriptions tend to revolve around the objects of the forest. Now close your eyes and continue your stroll. As you take a few more steps, other features of the world take hold: the exact positions of all the trees near your path, the slope of the trail, obstructive rocks, looming branches, and the sheen on the ground from raindrops that make the trail slippery. These features are totally absent from the conscious narrative of the world you were creating just a minute earlier. Some features, like slipperiness, aren't even objects. Most of these new elements are built from surfaces, textures, and edges that define action boundaries. Together, they set very hard limits on what you will be able to do. You can't walk through trees, your foot needs to be elevated over the rocks in the path, you'll want to avoid stepping on ice, and your head needs to duck below a limb. These surfaces are essential features that anyone must sense if they want to move through any landscape.

Eyes open, you can run down the trail and, with little awareness or effort, readily maneuver through an endless array of surfaces requiring specific actions—for instance, hopping atop rocks or lunging through deep snow. The surfaces constantly require on-the-fly adjustments so your feet end up in the right spots. This unconscious side of sensing, the one that takes in features in the world to allow you to move through it or on it, constitutes a fundamental kind of "action perception."* The philosopher Jakob von Uexküll was the first to recognize the close connection between what is perceived and what is possible when, in 1909, he coined the term "*Umwelt.*"[2] He realized that space itself is weighted with meaning that profoundly shapes an organism's behavior. Both a person and a squirrel will track spatial distances and textures as they navigate through the world, but they construct a profoundly different sense of meaning from those features, leading to a very different understanding of what their physical environment is.

In the 1970s J. J. Gibson was one of the first psychologists to point out that our environment is rich with action-relevant properties, readily detected by our sensory systems.[3] On a walk through the redwoods, a tree trunk across the trail is as solid and impenetrable as a brick wall. There is no decision-making about whether it is possible to walk through a tree or a brick wall, and human vision rarely makes mistakes about classifying a redwood as a barrier to action. The kind of perception that

*There are rare literary attempts to capture what action perception might be like. In the book *Flatland,* Edwin Abbott built an entire narrative around what one can see when space is collapsed down to two dimensions. He realized that the ease with which an individual moves in this flat world is predicated on the shape of a person in front of them. Some could be shaped like a point, others like a line. Both are easy to pass head-on, but the side of a line, a circle or square, would block your way. Remarkably, he had this prescient insight about similar hidden properties of the 3-D world back in 1884.

is used for action operates on many spatial scales. Slopes are sensed, cliff edges are avoided, objects are deemed graspable or not. A hiker will respond to a small rock in the trail up ahead as something that needs to be stepped over. No conscious mental gymnastics are needed to adjust walking cadence, to place, for instance, the lead foot at exactly the right distance from the rock. It just happens; the rock is sensed and refines the ongoing gait. Gibson was fascinated by this direct, unconscious process. It is profoundly different from the conscious experience and verbal descriptions of the world that we recognize through introspection and observation. Gibson's ideas emerged at a time when psychologists treated much behavior as a set of increasingly complicated reflexes, with animals just learning to make stereotypical and predictable responses to specific stimuli. Gibson tipped the logic of this by arguing that behavior is not just triggered by stimuli, it is constrained by stimuli. Seeing a redwood tree out in front doesn't make a person do a specific behavior. It prevents her from doing many alternate behaviors. Similarly, the gaps between trees all form opportunities for possible paths to take. In this light, almost everything in the physical environment provides relevant information for possible as well as impossible behavior.

Action perception is a deep, fundamental aspect of all organisms that are capable of movement, from the most complex vertebrates to some of the simplest of invertebrates. Take, for example, action perception in jellyfish. They typically have different types of eyes that likely serve different purposes, like helping them flee from a shadow that might be a predator or letting them move toward light. But that kind of stimulus-response behavior requires only the sensing of light, and there is nothing to suggest the creature has knowledge about the external world, a representation of surfaces or shapes that could influence its behavior. And then, there is the box jellyfish, *Tripedalia cystophora.* Unlike jellyfish living near sandy shores or open ocean, the box jellyfish lives in a complex maze of mangrove swamps. And unlike the others, the box jellyfish possess an impressive array of twenty-four eyes. These eyes are special as well because they have spherical lenses like those of a fish and sensors on the back of the eye, like a camera, all hooked up to a tangle of neurons forming an extremely rudimentary nervous system. To see what the jellyfish could do, Anders Garm in Lund, Sweden, made a swimming box for his jellyfish that had a steady current running through it.[4] When he put dark rods downstream in the water, the jellyfish handily avoided them by swimming upstream. But when Garm used clear rods, the box jellyfish bumped into the obstacles and pinballed through the gaps, proving they had used sight to evade the dark rods. By varying the width of the dark rods, he could measure the jellyfishes' visual acuity, roughly a finger width. He also tested a similar species, *Chiropsella bronzie,* which lives in a simpler natural environment consisting of a sandy bottom and far fewer underwater obstacles. Poor *Chiropsella:* it managed to avoid some of the visible black rods but with much less finesse than the mangrove-dwelling *Tripedalia,* repeatedly bashing into even the fattest obstacles. The difference between the two species of jellyfish reveals how direct sensing can drastically expand an organism's ability to act in an environment by converting what is sensed into a representation of what is possible and impossible. Whereas crude visual detection is sufficient for *Chiropsella,* whose few natural obstacles are easily spotted from a distance, the *Tripedalia* has transcended a life of simple stimulus-driven reflexes into one with elemental action perception, allowing it to move in a far more complex world.

—

After leaving the pack trail at the mountain pass, I walked across a wide meadow. The surface was covered in lush grass, smooth and easy to step through. It ended abruptly and I was forced onto enormous granite slabs, polished smooth by long-gone Ice Age glaciers. The slabs formed interlocking ramps leading up, down, and around one another. There was a bit of way-finding to get through, but mostly I enjoyed easy traversals across surfaces of every possible pitch and stickiness. Occasional brush and scraggly pine trees narrowed the path, forcing me to zigzag to maintain forward progress. I was cruising through these obstacles like *Tripedalia*.

While it feels relatively effortless to navigate through physical space, under the hood the brain has to do an enormous amount of computing to avoid crashing into surfaces. One way to get a sense of the heavy demands placed on unconscious perception is by measuring where a person looks as she walks. Who knows how much time is actually spent using your eyes to look at the boring stuff in the world: the path, the wall, the horizon? To answer this, Alan Kingstone and his students at the University of British Columbia in Vancouver asked participants to take a walk through the UBC campus while they wore an eye tracker attached to a headband.[5] The tracker detects where the eyes are pointing relative to the head. The key trick is that Kingstone also attaches a small video camera to the participant's head that records what the head is pointing at. By combining these two recordings he can see where the eyes are looking in the visual scene at any given moment. Throughout his career, Kingstone has made an eloquent case that if you want to understand what an organism is really doing, you need to study it in real-world environments rather than with a computer monitor in a laboratory. It's easy to imagine what the UBC campus might look like. It has the pleasant paths, foliage, and streets of a typical urban campus with lots of students walking to class. You might think you spend most of the time checking out other people. What Kingstone found was that walkers spend about 40 percent of the time looking at the path. Another 40 percent of the time is spent making eye movements toward objects, particularly if they are potential obstructions. Only about 20 percent of the viewing is directed toward other people, and these tend to be quick glances at the face or body. The implication is that under the hood, we have a highly developed system for directing vision to critical surfaces in the environment to guide our actions. The upshot of this is that the sensing and computational processing needed for even a rudimentary walk in the park are pretty demanding. On a college campus, 80 percent of viewing is devoted to mobility and only 20 percent to looking at other people. Complex wilderness settings might be even more demanding.* The fact that people can spend so much time looking at surfaces in their environment without any awareness that they are doing so is good support for Gibson's basic tenet: that there is a direct relation between what gets sensed and possible actions that doesn't involve fancy mental imaging, conscious awareness, or thinking.

Studies like Kingstone's reveal that underneath a person's conscious representations of things in the world—that is, all the stuff that is recognized and talked about, such as

*Novice hikers underestimate the amount of visual scanning required to navigate off of trails. The National Park Service and experienced mountain guides have many stories of hikers who fail to watch where they are going and who simply walk off a cliff. The theft of visual awareness by mobile phones has amplified the problem, with regular reports, particularly along the bluffs of the University of California campuses perched above the Pacific Ocean, of a texting student plunging to his or her death.

faces, furniture, houses, and cars—there lies a rich and hidden perceptual system necessary for moving through and manipulating that world. For most people, this hidden perceptual system for action is extremely difficult if not impossible to notice. Instead, people are consciously drawn to tangible features and objects. To get a sense of this bias, purchase some art supplies and try to paint an outdoor scene with oils and a brush. Novice art students will often focus on faces, objects, and bodies. The beginner is drawn to the consciously perceived representational matter that they access all the time. What is missing or messy in the painting is the in-between space and relative relationships across a scene. The beginner needs to be guided away from object-based features and toward what is in the remainder of the scene: the surfaces, the intrinsic geometric relationships, and the horizon.

Gibson introduced the term "affordance" to make the point that a person doesn't just react to surfaces.* Rather, the sensed surfaces not only determine what is impossible, they define what is possible. Affordances are the opportunities, derived from surfaces, that can be acted on. Actions and surfaces that go hand in glove. A chair in a dollhouse affords a place for a

* Since Gibson first introduced the term, much has been written by philosophers, psychologists, and cognitive scientists about what he meant by "affordances." Affordance as a concept has been repurposed in the domains of robotics, human factors engineering, and psychology. There are so many usages and meanings that it is easy to get sidetracked. Here I stick with what Gibson originally proposed: Affordances are *objectively measurable, latent (that is, "hidden") action possibilities* in the environment's objects and surfaces defined *in relation to the perceiver's physical capabilities* but *not dependent on the perceiver's awareness.* Affordances exist only if there is an actor capable of exploiting them. (Gibson, J. J. *The Ecological Approach to Visual Perception.* Hillsdale, NJ: Lawrence Erlbaum Assoc., 1979.)

doll to sit, but not for you. There are many wonderful examples of affordance that show how people can make extremely sophisticated perceptual judgments based on their own physical attributes. Without moving a muscle, a person can decide whether a barstool is low enough to sit on, a stair is too high to step up,[6] an object is within reach,[7] or a slot in a mailbox is of sufficient size to fit the hand through. The brain systems that calculate these affordances are exceptionally clever, flexible, and finely tuned to the physical geometry of both the external world and the body. They are also extremely dynamic and operate whether a person is sitting, walking, or running.

One of the most important visual cues for updating an affordance while a person is moving is "optic flow." Optic flow is the sense of motion one derives from changes in size, texture, or perspective. Brains use optic flow to update a geometric model of the surroundings that determines, in part, affordance. To get a sense of how this is demonstrated in the lab, imagine looking through a pair of virtual-reality goggles that situates you in front of the entrance to a virtual supermarket that has some automatic sliding glass doors that are slowly closing. Given how far you are from the virtual doors and your speed of walking, could you squeeze through the doors in time? If you are standing still without visual cues that provide optic flow, this is a much harder judgment to make. But if you are actually walking toward the virtual doors with cues, your ability to decide if you can make it in time is far more refined.[8] By using optic flow along with a sense of body motion, geometric properties in the world become much clearer. Affordances depend on both the way things appear in the world and the capabilities of the actor.

Evidence that the brain has dedicated areas for recognizing surfaces, obstructions, and affordances emerged when X-ray machines were first brought to the battlefields of France and Belgium in World War I, sometimes in horse-drawn wagons. As a young officer in the medical corps, the British neurologist Gordon M. Holmes had the brilliant idea of taking X-rays of the skulls of hundreds of wounded soldiers. From an X-ray, he could make a reasonable estimation of the probable path that a bullet took as it ripped through the head, and the likely areas of the brain that were damaged. It was grim logic, but it worked. His studies of battlefield casualties transformed neuroscience almost overnight. For the first time, it was possible to localize behavior to particular brain regions in living, albeit profoundly injured, people. Prior to that, the correlations were always done postmortem via a dissection of the deceased's brain. Holmes's X-rays led to a series of landmark papers, including seminal studies of changes to motor coordination after damage to the cerebellum, the organization of the visual map in the occipital cortex, and the location of brain regions that create a sense of space from 3-D images. He was also the first to show how vision might create affordances. He described in painstaking detail a series of six patients who had severe injuries to the visual system. Each had suffered from a bullet entering one side of the skull behind the ear and exiting through the other.[9] The patients had what was called "flat vision," with loss of depth and 3-D perspective.[10] What is more, there was something else fundamentally amiss. As Holmes put it, "How severely spatial orientation may be disturbed was seen in the difficulty which some of the patients experienced in making their way around simple obstacles. Case I, for instance, on running into a chair shuffled about in front of it, obviously confused and perplexed, moved toward one side and then back again, evidently failing to perceive that a single step would allow him to avoid it. Case II could not find his way back to bed when a small locker that he could easily pass around was in his path; and Case V had the same difficulty when he met any obstacle in the room." The injuries to Holmes's patients rendered them not much better at moving about than poor *Chiropsella bronzie,* the jellyfish that can't avoid obstacles. A century later, researchers are only just beginning to revisit how 3-D structure is formed in the brain by using state-of-the-art functional MRI scans. As healthy subjects view scenes, their visual areas make sense by relying the most on perceived distance and a sense of openness.[11]

By midafternoon I had worked my way across several miles of rock slabs, interspersed with an occasional icy tarn. The rock turned to sand and I ascended a shallow slope that ended abruptly. The far side was a cliff, dropping away for a few hundred feet. I lurched to a stop. The reflex to freeze at a cliff's edge is quite powerful. It can feel dizzying, as if gravity will suddenly shift and pull me sideways, launching me off the edge. My halt was abrupt and automatic, making me think it was an innate reflex, as deeply wired as the ganglion cells controlling *Tripedalia.*

Whether the perception of affordances is innate is an old question in developmental psychology, one that has been tested many ways with frustratingly ambiguous results. In early experiments, psychologists tested if this cliff-edge fear was innate by using human babies. It's an easy experiment that anyone can try with his own infant. The experimenter placed a wiggling baby on a long table that had a glass section built into its middle. The question was whether the baby would crawl across the transparent section, a perceptual cliff, to reach her beckoning parent on the far side of the table. Simple as this experiment is, the published literature on this cliff effect is an absolute mess of results. One could spend weeks adjudicating between authors who argued that cliff avoidance happens early and automatically as part of crawling, and others who claimed it emerges only with time and experience. The mess comes in part from the design of the experiment: each infant can be tested only once, and infants normally do a lot of random crawling. Thus the chances are pretty good that they will crawl away from an illusory cliff regardless of any innate sense. To get past this impasse, the question would eventually need to be posed in a different way and with experiments where babies or toddlers could be tested more than once.

Overcoming my frozen stance, I gingerly peered over the side, cautiously searching for a way to down-climb. Cutting back and forth across the cliff edge, I eventually found a sketchy path. It was a steep staircased gully filled with sand and small rocks, as slippery as a skating rink covered with ball bearings. Working my way down, I slid a few feet with each step. This could have been fun if it had been a ski slope or sand dune. But here, there were too many drop-offs and ledges to risk a long slide. As the steepness increased, my body position automatically adjusted. First, I descended by facing outward, but I started to slide. I switched to side step, but my feet still slid out of control. Then I faced inward, assuming the position one uses on a ladder gingerly stepping onto some rubble below. The rubble gave way and I whipped downward, grinding against the rock with my knees and elbows until I was finally stopped by a large rock ledge. Uninjured, I was nevertheless a bit worried about repeating the fall in the section below me, which looked even worse. I tried another technique. I descended feetfirst, but I also used my hands so I could easily sit on the ground like a crab on all fours. After three ineffective methods, this fourth style of descending finally matched the demands of the terrain perfectly. From the moment I peeked over the cliff edge to my final crablike descent, I had unconsciously tested a set of complementary actions to get down the slope, eventually finding one that would work. I was matching action and affordance in the most complex imaginable terrain.

Was I using innate actions or something acquired? To answer this, it helps to take a look at infants as they learn to move through the world. In one of the most entertaining research laboratories in the world, Karen Adolph has spent a career developing novel tasks that allow her to study how infants learn to move across surfaces and take risks. In an environment that is part gymnasium, part circus trapeze, Adolph videotapes crawling infants and barely walking toddlers as they try out a variety of physical challenges. In one, she made a ramp like a playground slide covered with carpet with a slope whose steepness she could adjust. The roughness of the carpet can provide friction so that an infant can easily crawl down it if it isn't too steep. Because infants love to crawl, they will repeatedly head down the slope. The critical research question is whether they will attempt to crawl down a slope that is too steep.[12] Could they recognize trouble? With parents at the ready, each child was placed on a platform at the top of the steep slope. He or she fearlessly proceeded, headfirst, off the edge and quickly lost control, only to be caught by a parent or lab assistant before getting hurt.

With years of experiments behind her, Adolph has proven that children are a blank slate as far as recognizing any potential danger associated with the steepness of the slope. Bad news for the innateness theory. The good news is that they very quickly learned which slopes were possible and impossible to navigate, and with this experience they would hesitate if challenged with a ramp whose steepness was in the narrow gray zone between possible and impossible.* From first encounters onward, it seems that affordances about slopes and surfaces must be acquired through direct physical experience. Adolph also tested if the knowledge a toddler gained from crawling on the ramp could be generalized to other actions, such as walking. To do this, she first had toddlers crawl on ramps of different steepness until they learned what was safe or impassable. She then placed them at the top of the platform on two feet rather than their bellies. If they had learned something general about a ramp's steepness, they would not be willing to walk down it. To everyone's amusement, the toddlers proceeded to walk down the ramp, without any recognition that it would be impossible to maintain control. Whatever they might have learned about the ramp while crawling had absolutely no influence on their decision to try to walk down the ramp. Here too, they eventually learned through repeated experiences what slopes were possible while walking, independent of what they had learned by crawling. This lack of generalization shows that most affordances are remarkably specific matches between a body and the environment, gained from direct physical experience.

One of the most important discoveries that Adolph made is that infants are little experimentalists who will try out all

*This need to learn by experience has posed unique problems for parents raising children among cliffs. In the nineteenth century, the families of Scottish lighthouse keepers settling on the barren Farallon Islands, located in the Pacific off of the Golden Gate, devised a simple scheme to keep their children from falling into the shark-infested waters below. They would tie their toddlers to ropes lashed across the ragged palisade of rock. A major rite of passage for their adolescent children was "going off the rope."

kinds of new physical techniques to expand their repertoire of affordances. A crawler who has learned that she can't make it headfirst down a ramp will try other actions to solve the problem. For example, she will flip around and try going down feetfirst. She quickly learns she can descend a steeper ramp by going feetfirst than by going headfirst. In a similar way, in the mountains I had learned a bag of physical tricks for getting down a rock chute without losing control. It was easy for me to take these affordances for granted, as I had no memory of ever having learned how to down-climb like this in the first place. It's easy to forget that I had learned them through trial and error. When I take city friends into the mountains I need to remember that many have never traveled on foot through complex terrain. They can have enormous difficulty wending their way across steep slopes or rough ground. They have little if any knowledge of what their body can do in a natural environment, no sense of how to move in this jiggered landscape. Clinging to bushes or rock walls, they don't know what is sticky or slippery, when a ramp is too steep, or how much they can trust their own feet to hold them up. Years of experience on these surfaces, experience that amounts to not much more than walking in a lot of weird places, allows me to see this terrain in an entirely different way.

Ironically, city friends who are clumsy at something as simple as walking in the wilderness are often superb athletes, many of them exceptional downhill skiers. They have learned a completely different space of affordances. Through extensive physical experience based on trial and error, they have implicit knowledge about the snow and ice surfaces that their skis will cut that the rest of us are oblivious to. Another sport that shapes affordances is parkour. This is a form of urban gymnastics where the athletes defy gravity by running up walls, vaulting over fences, and leaping between rooftops. It requires enormous physical agility, strength, and experience. Does the parkour athlete (referred to as a traceur) see the world differently from you or me? Jessica Witt and her students had the bright idea of asking them to make judgments about the height of walls.[13] To a traceur, a wall looks shorter than it does to a novice who has not learned to jump and vault over it.

Changes of action perception are not limited to walls and ski slopes. Even small objects can look different after physical experience with them. In rock climbing, the novice will try grabbing all sorts of features on a wall and through trial and error will learn what is good enough to hang on to. Expert climbers are most notable for their minimal movement. Each hand or foot is placed on a feature with certainty. Experts know it will hold their weight. They see and believe things that the novice cannot and, as a result, can readily visualize more routes up a wall.[14] With brain scanning, my laboratory captured differences in brain activity of climbers compared to nonclimbers, revealing how the visual pathways involved in sensing affordances can adapt with experience.[15] But it would be a mistake to think of an affordance as just a way that the perceptual system is tuned. It is much more than that. It is an intimate bonding of perception and action, a way of knowing what is possible.

Like learning to crawl and walk, learning to ski, parkour, and climb requires extensive physical experience leading to the knowledge of very specific affordances. Even so, there is no reason to think that some kinds of affordance can't be adjusted to almost instantaneously. For example, imagine you are staying at a high-rise hotel. A button on the wall of the hotel presents a unique opportunity to summon an elevator when it is pressed, but not to a small child or a wheelchair-bound adult who cannot reach it. Allowing the child to stand on a stool or giving the adult in the wheelchair a cane to extend his reach can instantly alter the action possibilities presented by that elevator button. Physical ability instantly changes perception of an action opportunity.

One way to test for rapid changes of affordance in the lab is by changing body size. In a study that should be exciting to anyone fond of disco dancing in platform shoes, Leonard Mark strapped wood blocks onto the soles of his participants' feet so their legs were instantly longer.[16] After wearing the blocks for just minutes the participants were able to judge accurately whether they were tall enough to climb onto or sit on high benches, despite the fact that they were very poor at guessing how tall the actual blocks on their shoes were. Within this one experiment, the geometry of the world was instantly transformed. This idea can be taken further by asking how perception changes with novel tools. Normally, our sense of three-dimensional depth and ability to estimate the size of an object works best for things placed at arm's length.[17] What happens when a person uses a tool like grabbers or long-nosed pliers that elongate his arm? Remarkably, his judgments about what is reachable change almost immediately.[18] Another dramatic example of rapid change was discovered in an experiment with pregnant women. Despite being very pregnant, these women were able to make accurate judgments about how well they would fit through narrow doors without bumping into the sides, consistent with the acquisition of a new affordance. One might assume that this ability had emerged slowly as their body changed its shape over eight months. The surprising result came from their control group. These non-pregnant participants strapped on a belly pack that simulated the protuberant abdominal bulk of pregnancy. Just like the pregnant women, they too could readily learn to fit through narrow doors, only they had learned to refine their affordance almost instantly.*

After getting down the steep slope, I reached a broad talus field. Talus is the rubble that forms at the base of mountains as they slowly erode and fall apart. It is an endless pile of sharply carved boulders, ranging in size from a large beach ball to a minivan and unquestionably the most difficult dry terrain to move over. It is the ultimate test of a person's locomotive ability. The only way to make progress was to hop from one rock to the next. This got exciting when I would land on a loose rock that forced me to react quickly to maintain balance before jumping to the next. The secret to this game is predicting which boulders won't roll and planning two or three steps ahead. My pack was starting to get heavy, making the talus jumping difficult. I fantasized about a pack mule to take my load, but a mule couldn't manage this chopped stretch of rubble.

For decades the U.S. Army has been trying to build a robot that can carry a heavy load across really rough terrain like talus. The modern infantryman has as much as 110 pounds of gear slowing his forward movement and limiting the ground he can cross. Serious money has been committed to the design of artificial mules, and the research is starting to bear fruit. Early efforts to make a mule, from the 1950s onward, focused primarily on the new kinds of mathematics needed to adjust movement based on feedback. Basically, key algorithms had to

*The speed with which an affordance can be updated likely depends in part on seeing one's own body. For example, it takes a long time for people to be aware of their enlarged size when wearing a backpack. To experience this firsthand, sit in an aisle seat of an airplane while it is being loaded with passengers and count how many times you are hit by a pack.

be developed to keep a robot from falling over, a rocket on a straight path, or a self-driving car from crashing. With elegant mathematical theory and fast computers, this kind of control has largely been attained. Given the successes in building control systems, why don't we see more robots roaming the Earth? That would require solutions to two difficult robotics problems. The first is a deep engineering challenge involving the integration of new materials and biomechanics to make energy-efficient artificial limbs that are sufficiently robust yet supple enough to move on different kinds of surfaces. A lot has been done to overcome this challenge. For example, there is a robot called Cheetah that weighs the same as a cheetah, runs as fast as a cheetah, and uses the same amount of energy as a cheetah.[19] The second hurdle is much harder: How do you build an artificial brain that can understand and learn affordances as well as the human brain? There is a huge gap in understanding that is only now beginning to be filled. Consider what has been missing from the robots of the past decade. One, aptly named "BigDog," is a 240-pound system built by Boston Dynamics that appeared around 2006. BigDog is capable of carrying approximately 100 pounds over a variety of rough terrains and light snow.[20] The walking problem for quadrupeds like BigDog was solved by incorporating goatlike motion in the legs, which relies on a fair amount of compliance—that is, springiness. Motors and cabling move BigDog's body with a vertical bouncing motion, similar to the hopping gait of a goat or trotting horse. For the robot to remain upright, the torques at the hips must be constantly adjusted and the feet placed at locations that assure a balanced symmetry. At first glance, the ability of BigDog to roam through a snowy forest is impressive. BigDog is incredibly stable, capable of steadily grinding up and down dirt piles. Look more closely, however, and many of the things that even a puppy or toddler can do are missing. The puppy learns to makes choices about the path it will take as well as where it will place each foot. On a talus field, it would learn to hop or bound from rock to rock. BigDog puts its feet at locations that move it forward and stabilize it, but not necessarily the best places. In a talus field, it could easily get its paws caught in the gaps between rocks. This is because BigDog is essentially blind. While it does have a few optic sensors to prevent it from smashing into trees or walls, akin to the big-picture affordances available to a box jellyfish, it can't see what's on the ground. To get a sense of what this is like, try walking outside in a park with your eyes shut, with a friend giving directions to help you dodge trees or barriers. To avoid tripping you will automatically lift your feet higher than normal and use a springy, wide-based walk to absorb surprises while maintaining stability, just as BigDog is designed to do.

In the decade since BigDog appeared, the design of actuators, motors, and springy robot legs improved dramatically, and bipedal humanoid robots became a reality. Their control systems are refined enough for them to jump, bound, leap, crawl, and sidestep.[21] They are compact, quick, and extremely energy efficient, like Cheetah.[22] These feats demonstrate that the biologically inspired artificial leg has reached a milestone of technical maturity. The problem is, most of the robots still can't see the ground out in front. They have laser vision that detects large regular surfaces like walls and steps. Otherwise, they have only a rudimentary sense of affordances. A talus field would quickly overwhelm them. And so, the next big challenge is to find ways of merging what is seen and where to walk: to create artificial affordances. This is hard because no two surfaces are alike, and so it can be difficult to find general rules for behavior based on what is seen. There are some amazing interim solutions. As Cheetah bounds along a treadmill it is able to adjust its stride so that it can leap over a tall foam block thrown in its path. This calculation requires not only that the front leg get over in time, but that the leap is long enough that the back legs make it over the block as well. The robot Little-Dog, which is the size and weight of a petite Boston terrier, is capable of adjusting its gait when it perceives rubber tile, rubber chips, small rocks, and large rocks.[23] Critically, it will adjust each step so that it places its paws on the tops of rocks along its path, just as I would do running across a talus field. The computations and sensing are still very slow, but it is only a matter of time before a robot will be able to create artificial affordances in near real time, allowing it to run through a talus field like I can, bounding from one rock to the next. The solution will likely require the creation of artificial intelligence endowed with learning algorithms so that a robot can build affordances from scratch, after the fashion of humans, who learn through intrinsic desire, trial, and error. And if this kind of robotic learning is anything like human learning, it will require a lot of training. A young toddler walks 2,368 steps and experiences seventeen falls per hour as he or she tests out the environment and learns what is possible.[24] One can envision a similar experiential process for training new robots to ambulate through increasingly complex situations.

It had been a long day of walking and climbing and my legs were tired. With fatigue came stiffness. Ahead I could see a small lake, just before a large drop-off. It would be a perfect place to camp, with an enormously wide view down a U-shaped glacier-carved valley. The ground leveled to a grassy meadow darkening in the shade of the setting sun. The grass was thick and deep but smooth compared to a talus field. I picked up my pace on what seemed like easy terrain and started to scan for campsites, ignoring the ground. And then I tripped and fell hard, flat on my face, the weight of the pack driving me forward. Hidden under the green verdure were rocks and holes, ready to catch the unwary. Fortunately, the grass was soft enough that nothing shattered except my pride. Sore, I rolled over in the grass and stared up at the sky, realizing I had gone far enough.

Doctors have an unglamorous name for my accident: a "ground-level fall." It is far less dramatic than an above-ground-level fall, which occurs when a person falls through the air, off a ladder, or down a flight of stairs. Ground-level falls, of whatever cause, are the number-one reason why a person ends up in a trauma center.[25] There is no end to the multiplicity of hips, skulls, and forearms that are broken in ground-level falls. Consider the distal radius, one of the two forearm bones near the wrist. Each day two hundred and thirty-five are fractured among the over-sixty-five U.S. Medicare crowd.[26] Recent estimates from the same cohort suggest that a bone is broken in only one-tenth of ground-level falls, giving an indication of how hard it is for older people not to fall down at some point. The causes of ground-level falls vary, with most of the medical field focusing on declining vision, impaired balance, and diminished strength with aging. These pernicious declines undermine a person's willingness to walk, particularly outside the home, further deconditioning them and exacerbating the risk of a fall. So it has been that the main strategies to combat falls have been cataract surgery, balance training, and strengthening exercises, all of which lead to a little less falling down in the elderly.[27] But there is a new movement in physical therapy and rehabilitation medicine that thinks this is not good enough. There are simply too many people still falling down who have good vision, balance, and strength. There is a new view that asks whether a person at risk of falling still has knowledge of her own affordances—that is, knowing what she can or cannot do. On the positive side, when an individual's physical capacity is diminished because of old age, there can be a well-calibrated diminution in her perceptions of action possibilities. For example, as people grow old, their estimation of whether they can make it up a flight of stairs is reduced.[28] Fortunately, in these cases, the action perception and physical ability are well matched. However, there can also be dramatic mismatches between what a person thinks she can do physically and what her aged body is actually capable of. This disconnect between self and the world is a kind of disembodiment, with a person becoming a brain in a jar, as it were.[29] Like one of Karen Adolph's toddlers, older adults overestimate their ability to stand on an incline or to clear an obstacle. They also fall down stairs and off ladders more frequently.[30] To counteract this mismatch, there is increasing recognition that lifelong physical experience is essential to maintain an intuition of what is possible and that one of the best ways to achieve this is by regular walking in complex terrains. Walking out of doors, with the attendant complexity of movement over varied surfaces, remains one of the most effective methods for reducing the risk of falls from aging.[31]

It is tempting to think that ground-level falls are a plight of only the elderly. But more than a fifth of adult patients showing up in the emergency room with a ground-level fall are younger than sixty.[32] Something besides age is making these otherwise healthy youngsters hit the ground. As with the Medicare crowd, the problem of course is not necessarily vision, balance, or strength. It is ultimately a problem of prediction, of matching the right kinds of actions with surfaces that present themselves: a problem of affordance. Falling can occur whenever anyone tries to generalize affordances when they really shouldn't, just as a naïve toddler does when placed at the top of a steep slide.[33] In my trek across the grass, I had made a prediction based on past experience about the smoothness of what looked like a level surface when underneath it was hiding a hodgepodge of jagged rocks. Around the country, the consequence of overgeneralizing affordance appears every midwinter as the surfaces get slippery.[34]* After every ice storm, the number of calls to emergency medical services nationwide jumps by a third. During the winter months, fracture rates reliably climb by a third. When questioned, most of the slippers admit to seeing the ice in front of them but wrongly assumed the surface was a familiar affordance that they can handle.

The epidemic of ground-level falls reveals how detached people can become from some of the most elemental affordances of the natural world. Living indoors, testing the elements only intermittently, people quickly become rusty at moving through natural settings. The affordances of nature are singular; they demand the kind of practice that cannot be replaced by the "physical fitness" achieved in gymnasiums. Fortunately, all of us are endowed with a brain that can readily learn to recognize the action-relevant possibilities of the natural world.

* Slipperiness is sufficiently problematic that it has earned itself a rating scale. Go outside and score 1 point for each of the following items. If the score is above 4, there is a 20 percent greater chance you will break your arm. The risk continues to climb for more points.

Temperature ≤32°F (1 point)
Presence of rain/drizzle (1 point) or snow/freezing rain (2 points)
Snow/ice on ground at start of day (1 point)
Fresh rain (or snow equivalent) that day ≤0.01 inch (1 point) or >0.01 inch (2 points)
Wind speed ≥24 mph (1 point)
Total = slipperiness score (0–7)

3

Shaping the Self

*Omnia mutantur, nihil interit.**

—OVID[1]

I DECIDED TO TAKE what, from my map, looked like a simple little shortcut. My route was a beeline that passed directly over a ridge and down a forested slope. I could have used a winding fisherman's trail instead, but it backtracked four miles out of my way. The direct way looked easy: just head for a V-shaped slot cutting through the rocks and then descend into a valley beyond. It began with fast walking over smooth granite slabs, enticing me upward toward the slot. But once there, things got messy. Descending the other side, I would need to down-climb a short, but steep, cliff. There was no easy way down the cliff, but I was too stubborn to backtrack. To each side of the slot the cliffs grew to enormous heights, blocking any alternate routes. I stared down the drop-off and could make out a reasonable climbing route laced with some cracks and small ledges. It would be strenuous with a backpack, so I took the pack off and launched it over the edge. It landed with a dull thump, underscoring the finality of my decision and providing a preview of what it would sound like if I fell.

* "Everything changes, nothing perishes."

With that, I started to climb down. The first few moves were familiar, no more difficult than climbing down a tall extension ladder. Facing the rock, there were plenty of handholds and places to put my feet. I soon reached a ledge, but the footholds below seemed to disappear. There was one small ledge, the size of an iPhone, pretty far down, and I wasn't sure whether I could lower myself to reach it. It was like descending a ladder with three of the rungs missing in the middle. Would it be possible to reach the fourth rung? I didn't want to lower myself to a point where I was hanging on for dear life, with my feet dangling in midair. Making matters worse, there was a bulge in the rock wall. If I arched my back to get around the bulge, the reach with my feet would be shorter and my toes might not find the little ledge below. Missing it would not be a happy occurrence. The decision depended entirely on one simple bit of information. What was the length of my body when stretched out? At first, this might seem a triviality for my brain to know. But once a body starts moving around and changing shape, keeping track of where it is and how big it is quickly becomes a really hard problem to solve.

Body shape is squishy, and a person's sense of her own shape is subjective. People feel bigger with a backpack on, smaller in tight clothes, and distorted with a wide-brimmed hat. And squishiness can get out of hand. Consider the story of a forty-three-year-old woman who lived in the countryside outside of Leeds. A mother and housewife, she came to the hospital gripped by fear and anxiety as a wave of bizarre feelings overcame her. She thought she was losing her mind, telling the consulting psychiatrist at High Royds Hospital in Menston, West Yorkshire, that on some days she felt that her head was double its normal size but half its weight, like a blown-up party balloon. On other days, she felt she had shrunk to half her normal height. An hour after her shrinking she was usually stopped in her tracks with a severe headache that put her in bed for the rest of the day. It was 1953, at a time in medicine when psychiatrists typically interpreted the mind almost exclusively through a myopic Freudian lens. They would have been quick to diagnose neuroticism attributable to obsessions and anxiety. Fortunately for this patient, the consulting psychiatrist was John Todd, who thought differently.[2] Another of Todd's patients was a forty-year-old man with a recurring feeling that he was much taller or shorter than usual. Sometimes he felt that he was eight feet tall, but at other times he felt as though he had shrunk to a mere three feet. In addition, he was often conscious of a feeling that his head was "twice its normal size and as light as a feather" or that either of his arms was missing. Todd suspected that the reason his patients felt their bodies squished into weird sizes was because of a disruption in parts of their brains that are essential for creating an image or map of the body. He was quick to recognize that the cause was a complicated form of migraine headache and not neurosis. Although he could make no claims as to where in the brain the disruption might be taking place, Todd confidently named his patients' affliction the "Alice in Wonderland syndrome."* Todd also noticed that their body distortions could change their feeling of how they were moving. For one woman, whenever her feet felt like they were a yard long, she also felt like she was going up- or downhill,

*The allusion to *Alice's Adventures in Wonderland* is poignant because Lewis Carroll was known to suffer from severe migraine headaches and may have also suffered from distortions of his body during his attacks. Todd is exemplary as a psychiatrist of his time for seeking neurobiological explanations for psychiatric symptoms. He also had a flair for creative naming. He first described the Othello syndrome, the syndrome of Big Brother, and Lilliputian hallucinations, and he provided additional case descriptions or analyses of the doppelgänger effect, Capgras syndrome, real and imaginary hypophallism, the Cheshire Cat syndrome, and folie à deux (the transmittal of psychotic thought).

even when she was on flat ground. Thus Todd was linking the feelings that arise from motion with the feelings of the body itself.

Todd's realization that there was a close tie between the feeling of the body's posture or shape and movement had already been suggested in the early nineteenth century by the Scottish anatomist Charles Bell.[3] In Bell's detailed dissections of different mammals, including humans, he discovered that there were separate nerves for sensing that could be distinguished from nerves for driving muscular activity. Before Bell's studies, it was taken for granted that vision provided positional information about the shape of the body in any given posture. In his analyses of natural movement by blind people, he realized this couldn't be correct. They could readily maintain postures, balance, and poise without sight. Bell ran similar experiments on himself by closing his eyes and established that he too could move and maintain body position remarkably well. He argued that there must be another subtle form of sensing besides vision, one that tracks body position and, in addition, is not derived from feelings that come from skin touch, muscle fatigue, mental effort, or visceral pangs. Rather, these were feelings specific to the movement of muscles and angles of joints. Although he didn't know what particular sensory organs were operational, he recognized their central involvement in making movements and proposed that they should be treated as a sixth "muscle" sense. Bell's notion of a sixth sense reverberated through philosophical circles for many years, fueling debate about the dualism that arises between the subjective sense of conscious or willful effort and the seemingly paradoxical lack of conscious awareness of posture or movement. How could a person consciously and willfully move while being utterly unaware of her own body's movements?

The idea of a sixth sense solidified in the 1840s with the

advent of a new medical specialty—neurology. One of the clinical problems that preoccupied early neurologists was a malady sweeping through the parks and streets of Paris, one that was peculiar to upper-class gentlemen. These courtly fellows, typically refined and of noble bearing and accompanied by their mistresses, could be seen on their daily strolls throughout the city. Despite their well-groomed appearance, their gait was altogether odd. With each step, the knee would be lifted higher than normal, the foot dangling. Then the foot would flop to the ground. It was hard to imagine this was a new style or cultural fad. Something was wrong. It was not long before it was given a distinct name, "progressive locomotor ataxia," by a particularly gregarious young neurologist named Guillaume Duchenne.[4]* This was just one of many ways that a person's gait could be altered, and the clinical challenge for Duchenne and other neurologists was to understand how damage to different parts of the nervous system might result in different kinds of walking patterns. There was a critical need to take measurements of different patients, but this was all happening long before cinematography or motion-capture technology. One of Duchenne's contemporaries, Gilles de la Tourette, developed a clever low-tech solution for recording different gaits. He had his patients step barefoot into a pan of ink, from which they proceeded to walk down a long hallway lined with butcher paper. In his classic eighty-eight-page treatise on gaits, he systematically describes each kind of abnormal gait accompanied by little footprints running down the margins of his book.[5] The section on progressive locomotor ataxia

*Duchenne was notoriously eccentric, trolling around Parisian hospitals without invitation looking for interesting cases. He would hold impromptu clinical rounds with the local doctors, who eventually embraced his enthusiasm for the study of the nervous system as a distinct clinical specialty.

includes a patient's footprints with a normal stride length, but the patient's feet were widely spaced and the toes pointed too far outward, like a duck walk. And there was more: the distribution of ink within individual footprints was all wrong. The foot wasn't hitting the ground heel to toe, it came down at once, splatting the ink. Tourette and Duchenne had the deep insight to understand that most of what they were observing in this gait reflected unconscious compensatory strategies; the brain was finding an alternative way to walk and at the same time keep the body from falling over. The main problem was that their patients' feet tended to drop down. This was caused by the brain disease. By taking a high step, the tip of the toe was less likely to catch on things. This was compensation. The outward turn of the feet provided more stability. More compensation. This was one of the earliest situations where neurologists recognized that brains are incredibly clever at finding alternative means to overcome an impairment, all to accomplish goal-oriented actions. Even today, the clinician is faced with this challenge of distinguishing what is a primary loss of function and what is a compensatory strategy in almost every neurologic disease.

It would take twenty more years before the world's first expert in venereal disease, Jean Alfred Fournier, recognized that the degeneration in these patients with progressive locomotor ataxia was caused by an invasion of syphilis in the spinal cord, subsequently referred to as tabes dorsalis. Paris was undergoing an epidemic of neurosyphilis, and those who could afford a mistress readily fell prey to it.[6] The infection was causing degeneration of the columns of fibers running up the spinal cord that carry the sense of position and touch to the brain. Testing a patient was easy: the patient couldn't tell if his toe or foot was being moved up or down by the examiner. The patient was still sensitive to a pinprick or the temperature of a piece of hot or cold metal, making it obvious that position and movement really constituted a distinct sixth sense as Charles Bell had proposed. With the introduction of antibiotics, the gait described by Duchenne all but disappeared, and modern neurologists rarely see this unfortunate sequela of syphilis.

By 1900 the different sensory organs within the joints, ligaments, tendons, and muscles that provide information about body position had been identified through the microscope. The computations these sensors enable are complex. For example, a sensor in the Achilles tendon will be responsive to how stretched it is and give a sense of the position of the foot. But this tension could reflect either the angle of the ankle or the tightness of the gastrocnemius muscle attached to the tendon. To make sense of position, this evidence needs to be integrated with sensors in the muscles of the lower leg. Together, all this position signaling is combined and called "proprioception."[7] However, a sense of position for individual joints is just a starting point. What the brain really needs to do is keep track of the entire body all at once. This all-at-once problem was first recognized a century ago by two of England's greatest neurologists, Henry Head and Gordon M. Holmes,[8] the latter of whom acquired his famous X-rays on the western front.* They coined the term "body schema" to embrace the dynamic properties of the body map. Their definition remains salient today. "For this combined standard, against which all subsequent changes of posture are measured before they enter consciousness, we propose the word 'schema.' By means of perpetual alterations in position we are always building up a postural model of ourselves which constantly changes.

*Henry Head was lead editor of the journal *Brain,* not to be confused with Lord Brain, who later became head of *Brain* and wrote an admiring essay on Head published in *Brain.*

Every new posture or movement is recorded on this plastic schema, and the activity of the cortex brings every fresh group of sensations evoked by altered posture into relation with it. Immediate postural recognition follows as soon as the relation is complete."

In Head and Holmes's time, the body schema was largely an abstract concept, something that could be studied only indirectly by tracking the changes to the gait in patients with tabes dorsalis or similar diseases. The body schema remained hidden from direct study until it was recognized that position sense could be tricked in healthy people. One easy way to do this is with a high-powered and relatively fast 20 hertz vibrator. If the vibrator is pressed against a tendon—say, the biceps tendon at the elbow—the position sensors at just that site are completely overwhelmed. The brain interprets this as a relaxed biceps. Meanwhile, the receptors in the triceps tendon on the other side of the joint continue to fire. If the person closes his eyes so that vision isn't available to double-check limb position, then the brain can be tricked into feeling that the arm is straighter than it actually is. The perceived shift of joint angle can be huge, many dozens of degrees. And the sensors in almost any joint in the body, even the fingers and hands, can be disrupted by vibration. The German army thought that the impact of vibrations on position sense was serious enough that they conducted a study on the disruption of hand perception that might occur from climbing ropes with different types of braids. They were particularly worried about fast roping, the special operations technique where a warfighter quickly slides down a line hanging from a helicopter. They were concerned that if the soldier slid too fast, the intense vibrations from a braided rope would overstimulate the position sensors in the hands and they would reach the ground not knowing where their trigger finger was.[9] The results were compelling enough to recommend changes in the texture of fast ropes.

Confusing the position of a body part with vibration is a jumping-off point for understanding how the body schema might be designed. The vibrator trick can be taken a step further by creating sensory conflict between the sense of touch and the sense of position. In 1988, James Lackner reported an ingenious series of vibrator experiments that did something better than just disrupting a person's estimate of the angle of a joint like the elbow.[10] Blindfolded participants in his experiments were instructed to touch themselves at various places on their own body while he stimulated a tendon in the arm doing the touching. For example, while they touched their nose with their index finger, Lackner stimulated their biceps tendon at the elbow. Because they were blindfolded, they couldn't cross-check the position of their body. A participant's brain was now faced with competing evidence. On the one hand, the person had an existing body schema that included a familiar shape and size for the nose on the face. And the sense of touch at the tip of the nose reinforced where the nose was located. However, once the stimulation began, the person sensed that her arm was extended. Thus the position of the hand and fingertip was perceived as farther away from her face than before. This created a multisensory conflict, with touch and joint senses locating the tip of the nose in different places. A brain might deal with this conflict in a couple of different ways. It might use a winner-take-all sensing strategy. It could believe in the one sense with the best signal quality and ignore all the rest. In this case, the winner would likely be touch to the face, beating out the position sense of the arm, and the distorted arm would be ignored. Alternatively, it could accept both the touch and vibration evidence as true and track two points on a body map at the same time, one for the original nose position and another for the newly created nose location. One nose is in the original position sensed by touch and the other where the extended hand is out in front. In doing so, it might chop the body schema up, so that there could be two noses in different parts of space.

The brain doesn't do either of these things. What it does is much more interesting: it creates a perception that the nose is longer than it is. Affectionately known as the Pinocchio effect, the nose "feels" as if it stretches out in front so that the touched position matches the new perceived extended arm position. The brain is reconciling the error in the location of the confused arm with the touch of the nose by stretching the body map to fit all the evidence. A healthy person's brain will allow itself to have only one nose, one right hand, and so on at a time. There is no chopping of the body schema into multiple copies or parts to reconcile these errors of position. One might suspect that this is just a quirky experimental result. But Lackner repeated this illusion by stimulating all sorts of tendons in the legs, the hips, and the back. If the triceps tendon was stimulated so that the arm felt flexed, then the subject thought her nose was short or pushed into her face. If she touched the top of her head, it could be perceived as taller, like a conehead, or shorter, like a flattop. This is squishiness at its best. While these experimental results are quite bizarre, the illusions reveal a deeply elegant property of the body schema.* At any given moment, a healthy brain integrates all available sensory information to create a single connected and locally smooth map of the entire body.

*Illusions reveal the conditions where a highly optimized, robust, yet fragile nervous system begins to fail. Illusion analysis is useful in the same way that failure analysis reveals to the engineer the performance limitations of a man-made system. Human evolution required many design trade-offs that are revealed by illusions.

It maintains this holistic representation even in the face of radically conflicting evidence, and even if it must generate a morphologic property such as a cone-shaped head that the person knows consciously is ridiculous. One might think that our capacity for rational thought should override this incorrect model of body position. However, in a healthy brain, the formation of a unitary body map inevitably supersedes conscious interpretations. Just as Bell had suspected with his sixth sense and Head and Holmes noted in their definition of the body schema, the representation of the shape of the body—-*i.e.,* the physical self—is created before individual senses reach consciousness.

Lackner's experiments showed in beautiful detail how the brain places a premium on reconciling discrepancies between two of our senses, position and touch, to make a holistic representation of the body. But what about vision? Vision is so precise at locating positions in space that it might assume a particularly overbearing role in updating the body schema. This was dramatically revealed by the rubber-hand illusion, one of the great parlor tricks in cognitive psychology.[11] The participant sits at a table with his hand and forearm underneath the table surface and out of view. A rubber forearm with a hand is placed on top of the table and positioned so that it is close to where the real hand would be located if it was on top of the table. The illusionist strokes the rubber hand with a paintbrush, and the real hand under the table with a second brush. After a very short time, the participant perceives the rubber hand as his real hand. At the climax, the illusionist picks up a hammer and smacks the rubber hand, causing the participant to jump up and scream as if his own hand were injured. The effect works even when the illusionist is pretty sloppy and doesn't position the rubber hand over the real hand. However, it will not work if the rubber hand is on the other side of the room, or if it looks like it is attached to another person. The trick relies on the brain being faced with incongruent visual, tactile, and proprioceptive sensory evidence. The person will see one hand on a table and feel the position of another hand under the table. Which hand is his? If the seen hand is within reach, the brain defaults to assuming that it is the observer's hand, even though the participant knows consciously it is a fake. Psychologists refer to this type of illusion as an example of "visual capture."[12]

The inverse of this illusion, what might be called "visual abandonment," has been described by astronauts out in space. As they worked outside the Mir space station, they traveled five miles a second, orbiting the Earth every ninety minutes. With each sunset, the transition from brilliant sunlight to darkness is abrupt, and the darkness is absolute. There is no dust in the air for shadows to reflect any light. The shadows are pure black. The astronaut Dave Wolf described an episode on his first space walk in 1997 where the darkness moved across the walls of the Mir space station, suddenly enveloping his arm.[13] In that instant, he describes a feeling that his arm had been amputated; it was gone. Vision was calibrating the presence of the arm position and then establishing that the arm was no longer in existence. These fake arm and amputation experiences show how vision often takes a lead role over touch or position sensing in situations where the brain is reconciling multisensory conflicts in a winner-take-all strategy.

These many different experiments make a strong case that the brain evolved to create a single, continuous, and unitary body map. In doing so, it avoids enormous computational problems that arise whenever a person tries to move. Suppose you wanted to scratch your ear, but you had the sensation that your arm was simultaneously positioned at two locations.

Which of the two felt hands should do the scratching, and what direction should it move in? Automatically and unconsciously combining evidence to build only one body schema greatly simplifies both the selection of which limb to use and how to plan its movement. Nevertheless, it just seems weird that the brain isn't smart enough to recognize that the rubber hand is fake and simply ignore it. What is the evolutionary advantage of this perceptual error? One can make a strong case that in the early evolution of hominins, there must have been major advantages for survival to automatically assume that a visually detected (and back then, always real) hand in front of the body was one's own. A hominin needed to know exactly where her hand was and in a hurry. For example, swinging quickly from branch to branch high up in the forest requires high-fidelity knowledge of hand position, or a branch will be missed. The value of seeing one's own hand was shown in the 1990s by my colleague Michel Desmurget at an INSERM research unit in Lyon, France. He ran an extensive series of experiments that measured how well people could point to targets in the dark, without any vision of their own hand. He briefly turned a light on and off before movement started, so that subjects could momentarily see the position of their own hand in their peripheral vision. The brief sight of the hand automatically combined with the information from all of the other position sensors in the body to greatly increase their accuracy at pointing to a target in the dark.[14] While this calculation might not seem very important for someone typing at their desk, where the hands remain in a fixed position, for a hominin in combat or a hunter throwing a spear this enhanced spatial precision could be a matter of life or death. Faced with a constantly changing body configuration from rapid limb movements, automatic binding of the body schema to a seen hand, before the conscious deliberation of whose hand it is, provided a crucial computation for rapidly updating the body position. The downside of this automatic binding process is that the brain is vulnerable to unnatural stimuli like rubber hands that weren't around when our species emerged.

An awareness of body position and shape begins in early infancy, perhaps even prenatally as fetuses move about, sucking their thumbs.[15] It is awareness that is constantly refined through physical exploration.[16] Within a few months of birth, infants begin to observe their own moving hands.[17] If they are lying in the dark, they will move their arm so a hand is visible within a narrow cone of light.[18] They are slowly learning how to calibrate what they see and feel until finally, at about five months, it becomes possible to see them use vision to correct their hand position during movement.[19] But when they reach nine months, their dependence on continuous vision disappears.[20] They can plan a proper hand trajectory at the movement onset just as Desmurget's adults could. This is a big developmental milestone. It means the infant, now with a well-calibrated sense of body schema, is free to use his vision for other things.

The importance of proprioception, particularly its impact on the ability to move, becomes dramatically clear when the sixth sense is completely erased. At the age of nineteen, Ian Waterman was a healthy butcher in Jersey, off the coast of France, who had an unremarkable viral flu from which he recovered. But the virus misled his immune system into identifying many of his sensory nerves as foreign, as if they too were a virus to fight. His immune system rapidly went to work, eliminating all of the nerves for proprioception and touch below the neck. He went to bed feeling fine but on arising in the morning fell to the floor in a heap, unable to stand or control his limbs, having completely lost the sixth sense of his body in space. One of the first things he noticed was that the floor didn't feel hard. Hardness requires touch and a resistance to the body's position sense against gravity. But these were gone. Instead, all he felt was the coldness of the floor. He was unable to sit up or move, and doctors thought his illness was incurable, that he would be relegated to a lifetime of wheelchair dependence, like other patients who had suffered this kind of nerve damage. But then Ian did an amazing thing. It was hard, but he figured out how to sit up by using vision and intense willful effort. Once upright, he had to sustain visual focus and remind himself to stay upright or he would simply fall back on his pillow. Over a seventeen-month period of determined rehabilitation, he taught himself to use his arms and hands to sit up, to stand, and ultimately to walk by relying entirely on visual cues. Once he regained these abilities you might conclude that vision can in fact provide sufficient information to create a reasonable body schema. However, the neuropsychologist Jonathan Cole spent many years working with Ian and meticulously detailed the kinds of things that Ian had trouble doing, providing profound insight into what the sixth sense offers to the body schema.[21] For example, Ian compensated for a loss of position sense by walking just like the patients with progressive locomotor ataxia, with a wide-based, high-stepping duck walk with flapping feet. While he had learned to stay upright with his eyes closed, it was an extremely fragile stance. At the grocery store, if he didn't watch what he was doing while picking up a potato he would tip over from the added weight. His body had no way of knowing how to compensate for the potato's weight with subtle shifts of posture that the rest of us take for granted. He couldn't tell if things were light or heavy. He couldn't run, jump, dance, or move quickly. Ian Waterman made it crystal clear that while vision is helpful at calibrating the body schema, it is the sixth sense that ultimately allows us to move quickly, naturally, and without thought.

My pack was still in the dirt far below and I was imagining how far down I had to go to reach the little ledge located somewhere below my dangling foot. I could no longer see my foot or the little ledge; the bulge in the cliff blocked my downward view. I lowered myself, confident that I would reach it. I had been climbing a lot that summer and felt in tune, with a good sense of my own body and how I moved over rock walls. My foot made it to the ledge, I found a solid stance on it, and I was able to shimmy around the rocky bulge at my waist. I worked myself into a friendly crack with lots of holds and descended the rest of the way to my pack. I have lots of athletic friends who are stronger or fitter than I am, but they would have struggled with that one particular climbing move. They would have nervously guessed what their body could do, with feet flailing about, and they would have missed the little ledge just a few inches below their shoe. They would have lacked familiarity with their own body's shape. A little bit of experience might have made climbing easier for them.

What does experience actually do for the sixth sense or the body schema? One possibility is that experience changes the acuity of position sensing at each joint, like a thermostat in a house that is accurate to within 0.25 degrees instead of 1 degree. There are so many different types of sensory organs for detecting body shape, maybe they get tuned up as a person moves about in novel ways. If so, a person should have a better sense of movement across the limbs that are most active. There is an entire field of sports science devoted to this one idea. Take a blindfolded soccer player, sit him down on an exam table with his legs dangling. Would he be better at detecting when you straightened his knee just a bit than a nonplayer? For every scientific paper concluding that athletes in a particular sport are more sensitive to change in their joint angles, there are roughly two or three showing they aren't. When data are this messy, it is best to walk away and look for other explanations.*

Another way that experience might improve the sense of body position is by making nerves faster at transmitting position sense to the brain, providing a person with quickened and perhaps sharper responses. This idea took hold in the early twentieth century after the baseball slugger Ty Cobb had his nerve conduction velocities tested with one of the earliest tube amplifiers capable of this measurement. *The New York Times* lauded his lightning-fast nerves as critical to his success. Unfortunately, this kind of explanation eventually failed on deeper scrutiny. Once Cobb's height and body temperature are taken into account, his nerve speeds look pretty normal. With modern technology, there are no definitive studies showing that practice or being an athlete leads to faster nerve signals.

The last explanation for the beneficial effect of experience is that it enhances the way the sensory information is interpreted in the brain. If so, then the key question for neuroscientists was finding where this experiential tuning might take place in the brain. After passing through a couple of relay stations, position signals from different body parts reach the primary sensory cortex. For more than a half century, neurosurgeons have had the opportunity to electrically stimulate the primary sensory cortex while asking their awake patients where they feel tingling. Each patient's cortex has the lovely property of being organized into a crude map of the opposite side of

*Of course, having good senses to start with can make an enormous difference.

a much-distorted-looking body. The more sensitive a body part is (face, fingers, genitals) the bigger the area of the map. My colleague Jörn Diedrichsen wondered if there was any fine-grained organization in this sensory area. Using one of the most powerful functional MRI scanners in the world, he looked at the pattern of activity in the area of the cortex related to the hand and digits.[22] Remarkably, touch or movement of each digit elicits a separate pattern of activity. Each person has her own completely unique signatures for the different fingers. Turning this around, knowing an individual's different brain patterns allows one to make a very good guess of which digit was moved just by looking at the brain scan, a simple form of mind reading. Neurologists have known for a long time that this primary sensory cortex is needed to sort out information coming from different locations in the body. Damage to a part of this cortex from a small stroke results in a loss of deftness.* The clumsy fingers aren't weak. Rather, the patient can't move his individual fingers quickly and precisely, underscoring the importance of being able to isolate the body parts for particular movements.[23] It does not, however, really create an image of the whole body at once.

An amazing property of the primary sensory cortex is experience-dependent malleability. One of my patients had lost an arm in a lumber mill accident. Although his brain no longer received sensory cues from his missing arm, it continued to try to build a sensory map of it. Over time, the map slowly changed until he began to lose the ability to distinguish sensations of the virtual arm from those of his face. The cortical map of the fingers normally sits next to the face map. With the arm

*In keeping with a long-standing tradition of endowing clinical disorders with the most opaque terminology imaginable, Hugo Liepmann named this problem "limb-kinetic apraxia" in 1920.

lost, the face map began to encroach into the finger area. Thus touching his face could elicit feelings in his phantom limb. This is an extreme example of plasticity after the loss of a limb. We can ask if there is any analogous plasticity in intact people as they learn a new physical skill. With practice, would the primary sensory cortex get better at sorting out incoming signals into the different parts of the body? This question was first explored experimentally in the auditory cortex of hooting barn owls, who have an uncanny sense of hearing. After they learned to distinguish specific hoots, the neurons in the auditory cortex were found to be more sharply tuned to the relevant frequencies. Thus it was reasonable to think that human experts have sharper tuning of touch or movement for relevant body parts in the primary sensory cortex. This has been tested with brain scans of expert musicians like concert violinists and novices. The scans typically show that evoked responses from stimulation of the fingers are enlarged in expert musicians.[24] We still don't know if a similar sort of expansion occurs in athletes, but it is reasonable to assume they may undergo similar changes.

While practice can shape the ability of the primary sensory cortex to isolate movements of the body, it is also the case that too much practice can lead to trouble. I knew this firsthand. After I made it off the cliff, I decided to note the cliff's location on my map. I sat on my pack and slowly marked out its specific location. The more I tried to make my mark precisely, however, the tighter my forearm muscles became. My arm locked up with a deep, painful cramp, turning my fingers into a stiff claw. There it was again, my writer's cramp, perhaps aggravated by fatigue from all the climbing, but what triggered it was trying to perform an exacting action like pinpoint writing. My writer's cramp is a good example of what can go wrong when the primary sensory cortex does a bad job of sorting out inputs from different fingers or the forearm. Not being able to distinguish the different fingers, the commands to move them individually comingle and lead to cramping and stiffness, also known as "focal dystonia." But what made the maps of my fingers coalesce?

In the 1990s, Michael Merzenich attributed this phenomenon to excessive repetition of one kind of movement.[25] To show this, he recorded the activity of neurons in the sensory cortex of monkeys he had trained to repeatedly squeeze a handle with a strong grasp. They pumped their grip thousands of times, always in an identical manner. Sure enough, with this kind of overuse, individual neurons for touch in the sensory cortex lost their specificity. The area of the skin that they would respond to grew tenfold. Position sense was likely to be blurred as well. This research led Merzenich to a prediction that anyone who does too much of one thing could lose the fidelity of her sensory maps and be beset by focal dystonia, caused by the repetitiveness. A checker at the grocery store, endlessly dragging items across the bar-code scanner, can get it in his forearm. A violinist will get it in her left hand or bowing hand, depending on how she plays. Trumpeters can get it in their lips. And lots of us get it while writing. Brain scans of people with focal dystonia support this theory of comingled sensing. The practical implication is that experience reshapes how the brain sorts out information from the body, and there is a delicate balance where too much repetition of just one thing can lead to a disrupted map, not that dissimilar to what is observed in an amputee, whereas a variety of skilled actions leads to overall sharpening of the senses. This is one of many reasons why cross-training across multiple activities, sports, or musical instruments can keep a person from developing this severe performance disability.

The primary sensory cortex distinguishes where information is coming from in the body, but it doesn't put the information together into a complete body schema as Head and Holmes believed. To get to the schema, you have to move just a little bit farther back in the brain, to an area called the "superior parietal lobule." The SPL is remarkable in that it combines touch, proprioception, sound, and vision to build a map of the whole body. This mashing together of the senses can be seen when neurons in the brain of a monkey are recorded while the examiner tricks it with the rubber-hand illusion.[26] As the monkey is exposed to the rubber hand, these SPL neurons will discharge as a function of the angle between the seen and felt hand, tracking the disparity between the different sources of evidence that the body schema needs to reconcile. The SPL also shows more activity in fMRI scans of people when they are shown a picture of a hand and simply asked to decide if it is a right or left hand.[27] This hand discernment, like the rubber-hand illusion, elicits an automatic binding of what is seen and what is felt.[28]

The body schema can be disrupted in patients who had a stroke in the SPL. They can feel touch and distinguish joint position just fine, but they can't put this information together to get a bigger image from touch.[29] For example, if you draw a large number or letter on the palm of a blindfolded SPL stroke patient, he can't tell what it is. He lacks a mental sketch pad for the schema of the hand. If a coin is placed in his hand, he is perfectly deft at manipulating it. However, he can't use size—which requires a calibrated hand schema—to tell the difference between a penny and a dime. Instead, he will feel the edge of the coin to see if it is smooth or rough. The critical role of the body schema in making movements can be tested in monkeys by reversibly inhibiting the SPL with a local injection of drugs. When a small area of the SPL is shut off, the animals will no longer prefer to use the hand that is tracked by the SPL* or perform tasks requiring coordination of both hands.[30]

It is reasonable to assume that lifelong physical experience will improve the overall fidelity of one's body schema. A wonderful way to test this is to ask people to estimate the length of each of their arms. Which of your arms is longer? How much longer? In most people, the right arm is about 1 percent longer than the left.[31] To indirectly measure how long people think their arm is, Sally Linkenauger slowly lengthened a tape measure in front of her participants and asked them to stop her when they thought that the tape matched their true arm length.[32] If they were right-handed, the self-reported sense for their right arm length was very close to the true length. Their left arm, however, was felt to be about 5 percent shorter than the right—too short. For lefties, both arms were perceived to be very close to true arm lengths. Lefties are generally more ambidextrous than right-handers, and there was no corresponding left-longer-than-right bias. Right-handed people chronically underutilize their nondominant hand over the course of development, leading to a slight shrinkage of the left-arm body schema.

One would expect that expertise in dance, sports, music, fine arts, or any physically demanding craft might further tune the local features of the body schema. This tuning can be found in repositioning tests. Here, a blindfolded person's leg is straightened and then returned to a bent angle. The person

*The motor and sensory areas of each hemisphere control the limb on the opposite side of the body. Thus, right SPL lesions or inhibition will make an animal less likely to use its left hand.

must decide if it is at the same angle. Or she is asked to actively return the leg to a given position without looking. There are many examples showing that athletes are good at this, like tae kwon do practitioners, who show smaller errors in knee joint repositioning tests than people who don't practice martial arts.[33] Other activities that emphasize very specific stances and postures, like dance, will also lead to an enhanced ability to precisely control where a foot and knee are positioned, compared to activities like soccer, where there are no stereotyped postures.[34] Soccer players are refined in other ways, particularly in terms of balance and maintaining verticality across a variety of positions. The ability to integrate the body schema with verticality and balance also becomes invaluable in acrobatic sports like gymnastics and diving.[35] Nonacrobatic athletes tend to rely more on gravity to help judge their body shape, whereas gymnasts are better at judging their body shape in repositioning tests whether they are upside down or right side up.[36] No matter what the activity, there is opportunity for the schema to tune itself using any one of a number of senses.

Holmes and Head had noted that it might also be possible to investigate conceptual knowledge about the body schema by simply asking patients to point to named or touched parts of the body. They wondered if conscious knowledge of the body is one and the same as the unconscious body schema built by the sixth sense. A difference between these two kinds of body awareness was first observed in 1908 by the Czecho psychiatrist Arnold Pick. He described patients who, with language function, motor skills, and movement intact, were unable to follow simple commands such as "Touch your elbow."[37] He called this impairment "autotopagnosia," which is just a fancy way of saying ignorance of one's own body. What exactly did Pick's patients not know? Patients with this disorder are rare, but over the course of a century there have been sufficient new cases to shed some light. In 1991, while at the National Institutes of Health, the neuropsychologist Angela Sirigu described a sixty-two-year-old woman with dementia who also had a problem with naming body parts.[38] Sirigu designed an ingenious set of experiments to determine if the woman's autotopagnosia was from a loss of body position sense or a lack of general conceptual knowledge about body parts (ears are for hearing, feet are for walking). She needed to be certain that any errors made by the patient were not due to a basic loss of motor ability, speech, or vision. In one of her experiments, she taped small wood objects such as a toy car, a dollhouse bed, a rubber band, and a toy soldier onto different joints of the patient in clear view: on her knee, shoulder, sternum, elbow, and thigh. The patient had no problem when told to point at any of the named objects, such as the car taped to her shoulder. Sirigu next put similar objects on herself. Then she touched an object on the patient (such as the car). The patient was able to point to the same object on Sirigu with only rare mistakes. Thus, whether by verbal or nonverbal commands, she could identify three-dimensional objects and point accurately at a body. This shows that her ability to access her own body schema for the purpose of pointing was perfectly fine. However, when she was verbally asked to point at her own elbow (the same one that she had previously pointed to when a tiny bed was taped to it), she was lost. Or when her elbow (rather than an object) was touched, she had enormous trouble touching Sirigu's elbow. One common kind of error she made was intriguing. She often substituted a related body part (a knee instead of an elbow), suggesting some residual knowledge about what the functions or relative positions of those body parts might be. But she had lost any sort of mental representation of what a holistic body is like, irrespective of the names of the body parts.

This unique patient reveals how the brain stores knowledge

about a body in an entirely separate category from nonliving objects, and both of these kinds of knowledge are separable from the body schema for action. Follow-up experiments in a larger group of patients who had brain scans were needed to figure out which parts of the parietal lobe were needed for these different ways of knowing the body. One group of patients with parietal strokes had errors in a task closely related to the rubber-hand illusion and the updating of the body schema. Not surprisingly, they had lesions located in the superior parietal lobule. Another group failed tests of knowledge about the body such as locating or identifying body parts, akin to what Pick and Sirigu had described. Their lesions were located more laterally in the parietal cortex.[39] Thus the behavioral dissociation of these very different ways of knowing the body is reflected in very different corresponding brain areas. The brain is so designed that there are multiple bodies in it, a schema, an image, and a parts list, each existing in segregated computational layers. They perform profoundly different functions. A body schema enables movement for any possible mass or size no matter what a person consciously thinks he looks like or what his body parts are called. Putting this in more personal terms, all the naming of elbows or knees in the world wasn't going to help me get down the cliff. I needed a schema!

After getting off my cliff I looked forward to a pleasant but steep downhill hike through a forest to a creekside camp in the valley far below. However, a storm had hit the area the previous winter, and many of the trees, all massive old pines, had been knocked down, turning the slope into a maze of crisscrossing obstructions. I was doing well weaving between trees, my affordances of my body size closely matched to the widths between the trees. But then I came to a tree that was horizontal. I was quickly moving down the slope, sliding in sandy dirt. I crouched to duck under the log, not realizing that I had a pack bulging upward on my back. While I readily cleared the log, my pack didn't, and I was slammed downward onto my rear in a cloud of dust and dirt. Even though I had a great sixth sense of my movements that worked quite well for climbing and intact conceptual knowledge of my body parts, something was missing: my body image had failed to take into account what it looked like with a pack on.

Somewhere between conceptual knowledge about a body and the body schema lies self-image, what a person thinks she looks like at any given moment. When we look back on Todd's migraine patients with the Alice in Wonderland syndrome, we can see that they had normal knowledge about body parts, but their conscious self-image was profoundly distorted. Over the last two decades, many migraine patients have posted self-portraits, whether drawings or paintings, documenting alterations of their body image. The pictures show giant hands, extra limbs, or big heads.[40] We know that with complicated migraines there can be a wave of depressed cortical activity that sweeps across the surface of the brain like a slow-moving tsunami, disrupting anything in its path.[41] Todd's patients were likely experiencing a disruption sweeping through the parietal lobe and breaking the link between a well-calibrated body schema and a coherent image of the body. These distortions of body image are also reported by patients with epilepsy prior to the onset of a full-blown motor seizure. For example, Ronen Lecker, a neurologist at Hebrew University, described a fifty-eight-year-old woman who had an aura "during which she felt herself become smaller and smaller while no change appeared to happen in the surroundings," while feeling deep fear.[42] A brain perfusion scan showed a decrease of blood flow through the right parietal lobe. Disruption of the SPL by a seizure also fractured the normally holistic body schema. She described a transient sensation of being "split," accompanied by a hallucination of a second head. "Quite suddenly everything seems strange, and people's voices become very faint. I feel that my head is dividing into two. The second head seems to flow off my normal head, and to take up a position a little behind and to the right of it. This 'astral' head appears in the form of a vague, misty shape with a black outline. I feel that it is the detached head that contains my mind."

The idea that there can be a profound disconnection between the body schema and what a person creates as a body image helps to explain some of the scientific arguments surrounding a potential cause of eating disorders in young adults. Why would an otherwise healthy teenager starve in the face of an ever-shrinking body? Extreme conditions such as anorexia are driven by an extremely complex set of pathologic processes that include, but are not limited to, a potentially catastrophic disconnection of the body schema and the body image.[43] There is something profoundly altered in their body schema. Anorexia patients have misperceptions in touch. It takes more force for them to recognize that their abdomen is being pressed on[44] and they are less accurate at judging distances between different spots that were touched on their bodies.[45] These distortions lead to measurable changes in the way anorexia patients use their body schema to move in the world. When forced to squeeze through a narrow doorway, they unconsciously anticipate a squeeze by rotating their shoulders more than a healthy person of the same body size.[46] It is as if they are planning actions for a body that is bigger than theirs.

It is possible to have a profound distortion of just one part of the body image. There are people who spend their entire lives convinced that a body part, usually one of their legs, is not theirs. This has been labeled body integrity identity disorder, or BIID. The leg works perfectly well, but it doesn't feel normal to the patient. It feels like an appendage stuck onto them, there to do work like walking, but there is no sense of ownership. This feeling of alienation can be so strong that the person might seek to have her perfect-functioning leg amputated. Once the leg is removed, such patients express immediate, long-lasting relief. Why would anyone prefer a prosthesis over a good leg? The current thinking is that their parietal lobe has not developed normally, so these patients have not been able to link the sense of touch or position used for movement with an image of the body that includes that particular limb.[47] Without cross talk, the body image does not corroborate or confirm that the leg is actually part of the person.*

After knocking myself flat trying to duck under the log, I decided to name my problem "pack integrity identity disorder," a failure to incorporate my bulging backpack into my body schema. Without a sense of feeling from the pack, it was easy to ignore, and easy to underestimate how much it contributed to my body image.

Anorexia and body integrity identity disorder are extreme examples of something that exists in milder form in almost everyone. Everyone thinks he is too fat or too thin, much of this

*I continue to experience a mild version of BIID ever since I had a right total knee replacement. I am severely bowlegged (genu varum). The surgery corrected this on only one side, making my right leg perfectly straight. When I look down at my legs, however, I inevitably think my surgeon made a mistake. The new normal straight leg looks completely abnormal. It no longer matches my familiar crooked self-image, still preserved in my bowed left leg.

driven by social norms. One of my patients was a competitive bodybuilder, one of the strongest women I have ever met. Unfortunately, she began having double vision and balance problems, and after a detailed workup I made the diagnosis of multiple sclerosis. In counseling her, I was rather optimistic. The disease varies enormously in how it impacts each person, there are many sophisticated therapies for it, and she was very strong to start with. I was stunned when her immediate response was a desire to commit suicide. She said she would rather be dead than weak. Even though she hadn't even started to get weak, it was simply the threat of weakness that drove her determination. Despite a normal body schema and strength, her body image was already crumbling. Image played a dominant role in her personal identity, in defining who she was. It took me years to talk her down from the ledge, to convince her that she could have a meaningful identity irrespective of her body image. Fortunately, her disease progressed slowly, and with the help of intensive physical therapy and medications she maintained good strength. She is an extreme example. But a disconnection between self-image and actual body size also occurs in roughly one-quarter to one-half of all young adult athletes, in virtually every sport tested. Whether young athletes are driven toward a larger (football) or leaner (distance running, equestrian sports, gymnastics) body weight, distortions of body image can dominate their eating behavior, even when their actual body weight and musculature are at a performance optimum.[48] As with anorexia, the body self-image dominates the eating behavior. Why doesn't the athlete's well-calibrated body schema do better to control their eating behavior? Athletes are clearly sensitive to their own body mass and size when they play sports. Shouldn't this improve their body image? A simplistic explanation is that the unconscious body schema and the conscious body image are like two ships passing in the night. Each can influence the other, but only in fleeting moments.

I arrived at a quiet creek and found a small campsite, soft with pine needles and sheltered by a rock windbreak. Stiff from climbing, hiking, and ducking, I laid out my pad and flopped down to rest. My mind wandered, and I started to think about rodeo and that maybe I should do some yoga. I had recently volunteered as a doctor at a Professional Bull Riders event at our local rodeo. I knew from TV that bull riding can be rough. I got a completely different perspective at ringside. The bull and rider look like they are launched out of the chute like a symbiotic rocket, whipping about in multiple directions all at once. Riding a bull requires as much body awareness and dynamic strength as any sport out there. The ability to maintain form atop the bull, no matter what it does, is astounding. I spent time talking to the trainers, therapists, and riders. I wanted to know how they train in the off-season. How do they keep their edge without beating themselves up? They were already as strong as a steel cable. What they wanted to improve was their acrobatic performance. Remarkably, what they emphasized wasn't strength training or endurance. Rather, they focused on the need for better body awareness, and what many were doing to achieve that was a daily fitness regimen that included yoga.

4

The Hidden Hand

I'll consider it a win if it clears the pad and doesn't blow the pad to smithereens.

—ELON MUSK[1]

THE SKIES WERE overcast, with only a brief sunrise before heavy clouds swallowed up the light. I wandered down to a nearby section of a stream with a current that slowed as it spread into a broad pool. Down in the bottom, murky from the low light, swam a very large golden trout. I hadn't really brought enough food and was probably losing about a pound of body weight a day. The fish was a tempting supplement, but I hadn't brought a fishing pole. I fashioned a spear from a reasonably stiff branch. Wading into the icy shallows, I took my time easing closer to the trout. It was lazily holding itself against the current, occasionally letting itself glide. I stared at it for a while, trying to guess which direction it would dodge if it sensed my presence. I made a guess, but the fish swam in the direction opposite of where I thought it would go. I was pretty far away and the spear flew through the air into the stream, digging into the gravelly bottom and missing the trout by what seemed like a mile. This was going to be harder than I thought.

My directional control of the spear was poor. I needed to make some adjustments. First, I had to get closer to the fish. Second, I had to hold on to the spear longer, for then I would be able to sense the spear's likely accuracy and to refine my movement just before the spear left my hand. This is the essence of feedback control, the ability to sense and respond so that performance of any action mechanism, including my arm, can be improved.

Feedback control was first put into practice in the 1920s when mechanical contraptions were invented to automatically guide ships along a particular course. Despite their largely successful designs, these automatons were universally rejected by sailors worried about job security, not unlike taxi drivers worried about self-driving cars. With the outbreak of World War II, feedback control really took off as a way to improve the aim of weapons systems, including antiaircraft guns, torpedoes, and, by the end of the war, rockets. Before the war, rockets were still being guided like a child's toy: point it, launch it, and get out of the way. The technologies that emerged to control weapons like the V-2 rocket changed everything. The V-2 was a self-guided unmanned missile, launched from the Netherlands, which could cross the English Channel to strike anywhere in southern England. Even though the payload was small, the psychological impact of this seemingly unstoppable weapon was terrifying. Before the war, Wernher von Braun and his team of engineers in Germany had spent a frustrating decade trying to control a rocket's trajectory. In the process, they crashed hundreds of prototypes and blew up a lot of launchpads. Eventually, they discovered that if they used simple feedback control, similar to the automatic guidance of ships, they could roughly sustain their rockets' trajectory toward London. By measuring the difference between a rocket's actual and desired direction they could adjust its fins with little electric motors to make crude, on-the-fly corrections. But too much correction resulted in too much counter-correction and vice versa, resulting in oscillation. They needed to be able to control the sensitivity of the system. Making matters worse, the rockets were top-heavy and loaded with fuel, and any wobble at takeoff would lead to particularly severe oscillations. Without adequate control, they could flip and fall out of the sky, crashing explosively. To solve this problem von Braun used two gyroscopes: one tracked the true "up" direction and the other the direction of the rocket's nose. A simple algorithm measured the difference between the two and the output was used to drive the motors of the rocket fins. It took years to fine-tune the control algorithm so that the relationship between what was sensed as an error and how much to correct for that error was just right. This kind of control is also what it takes to stand and hold oneself upright. Overcorrect (think of a drunkard) and you will sway all over the place.

Such sensitivity adjustments were also in play as I practiced throwing the spear. To gain better control I would need to lunge and release the spear close to the water so it would stay in my hands as long as possible. I worked at it until I was able consistently to hit a target rock at the bottom of the stream. I was ready.

I stood and waited until a trout appeared. It was not as big as the first, but it was legal size. I stared at it, watching it lazily slip through the water. Unlike my target rock, it never stopped moving. It knew how to glide in the slow current, how to use the current to amplify its actions. This was going to be hard! I needed to speed up my feedback control, to find a way to adjust my movements midthrow to account for whatever changes the fish made in direction. Mine was similar to the problem von Braun faced. He could get his rockets off the launchpad and headed toward London, but the feedback control was too slow to make quick adjustments to changing wind patterns and minor perturbations such as buffeting from clouds. Without better ways to detect small changes in position or to anticipate small changes of a rocket's direction, von Braun was stuck with rockets that landed all over the English countryside as much as they did in Central London. He would only achieve reasonable steering many years later, with more sophisticated versions of feedback control that ultimately got U.S. rockets to the moon. One of the tricks for overcoming von Braun's problem in correcting his rocket's flight path was the creation of an internal signal that could be used to better anticipate the wind and buffets. For both trout and man, there is an analogous internal signal created by the brain that indicates ongoing and expected body motion and that can be incorporated into feedback control. For man, it is like a hidden hand, for the trout like a hidden fin: a secret signal that provides insider information about what the body is up to.

The hidden hand of action is best revealed in patients who suffer from a rare syndrome after sustaining a stroke that results from a unique pattern of brain damage. One such patient was an elderly woman who was sitting at home alone, busy knitting a sweater for her grandson, when a sense of dizziness swept over her. She dropped the knitting needle from her left hand and promptly blacked out. Her son found her on the floor later that day and rushed her to the hospital. With nourishment and care she soon awoke, but it was clear that she had suffered a stroke, leaving her with profound weakness on her left side. She was as chatty as ever and didn't seem to really care all that much about the weakness. When asked about her left side and how it felt, she was indifferent. She ignored her unmoving left hand. When it was picked up and shown to her she said it didn't really belong to her anymore. Clearly, she had a neglect syndrome, and there was a black hole within her mind that had swallowed her ability to recognize part of space. But there was more. The next day, she started noticing something strange that she was reluctant to share with her family or doctors. She felt and also saw a new, healthy arm attached to her left shoulder, in front of the old "dead" one. Despite the paralysis of her old left hand, she thought it was high time for her to finish her knitting project with the aid of this new hand. She could feel and see it, right in front of her. She became increasingly frustrated because the new healthy left hand kept dropping the knitting needle that her right hand passed to it. Her family also noticed her right hand repeatedly dropping knitting needles into her own lap. She reluctantly confided in them that she now had three hands, but unfortunately the new one couldn't knit any better than that old "dead" hand underneath.

MacDonald Critchley first described a case like this in 1953 and diagnosed his patient with a tongue twister of a syndrome: "supernumerary limb."[2] At the core of his and subsequent descriptions is a virtual hidden hand invented by the brain that creates very detailed and plausible sensations of what the patient attempts to do, such as a goal-oriented action like knitting. The hidden-hand phenomenon reflects the remarkable capacity of the motor areas of the brain to predict what the motor areas expect from the muscles involved in a movement. For the elderly patient, the signals from her motor areas no longer reached the muscles of the arm or hand. Nevertheless, these motor commands were still being sensed by the rest of her brain, areas unaffected by the stroke.[3] Even in the absence of real movements, the knitting grandmother consciously experienced a virtual sense of real, goal-directed movements. The paralyzed arm was lost in a black hole of neglect. Seeking contiguity in the body schema, her brain created a new arm to reconcile mismatches between what her motor areas were predicting she ought to feel and any conflicting sensory information. The resulting change to her body schema, with the inclusion of a new functioning arm, is a breathtaking illusion. Not only does she feel movement, she actually sees a body part that corresponds to these movements.* Admittedly she is a wild example of the distortions of reality that can occur with brain damage and a hidden hand run amok. But her case also poses important questions about the nature of the motor signals that create a hidden hand.

Most of us don't need an extra limb, so what useful purpose does this internal hidden hand provide for us? It is very hard to study these covert signals in the healthy brain. Nevertheless, a steady stream of brilliant behavioral experiments reveals how this signal can be used to solve the two problems of movement that the rocket scientist, trout, and I struggled with. The first problem that the hidden motor signal solves (whether for a hidden hand, head, leg, fin, or body) has a close analogy in Einstein's special theory of relativity. Einstein, sitting in a train, observed another train on the adjacent track passing by. Without any background information, it was impossible to tell if his train, the other train, or both were moving along the tracks. Similarly, how did the fish I was trying to impale know whether it was moving itself or being moved through the water?

In the late 1940s, at a pleasant marine science laboratory in Bimini, an island southeast of Miami, Roger Sperry began doing strange things to the eyes of southern swellfish and watching for changes in their swimming behavior.[4] He would surgically switch their eyeballs, so the left became the right and vice versa. Unlike those of humans and other mammals, the optic nerves of swellfish will regenerate and restore vision to some degree. Sperry realized that fish too must solve the

*Even though vision is the dominant sensory system, the patient was capable of overriding visual reality to create a false mental image of a new arm in order to preserve a holistic sensorimotor experience.

relativity problem. In its simplest form, the question is how can a fish tell if it is swimming forward or if water is simply moving past it? There might be a visual signal indicating movement in either case, but it won't clarify whether the fish is swimming or being moved by the current. Vision by itself can't solve the problem. So Sperry sought to understand how fish combined vision and the motor signals that make it move. If it was actively swimming, then visual perception of motion needed to take into account self-propelled movements. Sperry's surgical eyeball switch always resulted in the same pattern of behavior: the fish swam in circles. Sperry deduced that the motor areas in the fish's brain send out a signal to other parts that correspond to anticipated swimming direction. He called this a "corollary discharge." Under normal conditions, this discharge was matched with motion detected by the eyes, creating a concordant mapping of predicted physical and sensed ocular motion. When the eyes were flipped, there was a mismatch, leading the brain to sense too much motion on one side and not enough on the other. To balance things out, the fish swam in a circle. He concluded that "a corollary discharge of motor patterns into the sensorium may play an important adjustor role in the visual perception of movement along with non-retinal kinesthetic and postural influences from the periphery." Those last two bits are, of course, the sixth sense.

Sperry wasn't alone in recognizing the importance of this hidden signal of movement. In related work, Erich Walther von Holst and Horst Mittelstaedt also considered this kind of experiment, but they viewed the problem as being one of perception rather than action.[5] For them, the question was how to improve perception in an organism that is already moving. Just as a photo will be blurred if you jerk your camera, a visual image might get blurred whenever the eyes move. To study this problem, they flipped the sensory organs on insects instead of fish. Like Sperry, they observed circling behavior and recognized that the animal generated an internal signal that they called "*Efferenzkopie*," now referred to in English as "efference copy." They thought that this copy of the motor command was used to cancel out any sensory input whenever there was self-induced motion, like a camera that took a picture only if it had motion detectors confirming it was stationary. This cancellation idea was reminiscent of an experiment proposed by Hermann von Helmholtz in the 1850s.[6] He pushed on his own eyeball and noticed that the images appeared to move. However, when he moved his eyes voluntarily, he didn't notice any such effect. He thought that the motor command to move the eyes, what he referred to as the "effort of will," must interact with vision to create visual constancy.* In subsequent work, Mel Goodale and Claude Prablanc in Lyon, France, were able to show how powerful the efference copy is in the vison of people.[7] They had participants stare at a small dot of light about two feet out in front of them in what was otherwise total darkness. The light turned off, and at the same time another one turned on about a foot to the right of the first. The participants' task was to look at the new target. The eyes left the starting point and headed off for the new target. The investigators carefully tracked the participants' eye movements, and once the eyes surpassed a velocity threshold Goodale and Prablanc turned off the second target and turned on a third one, a little bit farther to the right. Thus, the participants' motor command wasn't strong enough for the new location, and their eyes would come up short. With

*While there is good evidence that there is a corollary discharge that underlies this effort of will, it is also apparent that the creation of visual constancy is extremely complicated and the corollary discharge by itself is too slow, too weak, and too inconsistent to maintain a constant perception of visual space by itself. See Bridgeman, B. "Efference copy and its limitations." *Comput Biol Med* 37 (2007): 924–29.

that, they would make an additional corrective eye movement finally to reach the new location. The amazing thing about this experiment was that the participants had no idea that the target had jumped. Nor were they aware they had made the extra corrective jump to the new target. During the initial eye movement, they were effectively blind. The efference copy was shutting off input, solving Helmholtz's quandary as to why we don't see the world move when we move our eyes: we are blind in that tiny moment of time when the eyes jump from scene to scene. From Helmholtz, Sperry, von Holst, and Prablanc we begin to see concrete evidence that there is in fact some sort of internal hidden signal coming from the motor areas and that one of its functions is to cancel out the sensory consequences of self-induced motion, yielding refinements of perception and better predictions of how one is moving.

It is reasonable to presume that the efference copy might also play an important role in the interpretation of other kinds of perception besides vision, such as touch. Whether you rub your palm across the corner of your desk or have someone run a blunt pencil across your palm without your moving, the stimulation to the touch receptors of your hand is roughly equivalent. How can your brain sort who is doing the touching? Again, the efference copy offers a framework for resolving this uncertainty. While a postdoctoral fellow at University College London, Sarah-Jayne Blakemore showed this by posing a simple question: Why can't you tickle yourself? To answer it, she resorted to an experiment with a paintbrush and a simple robot arm. If her participants rubbed a paintbrush on their own palm they felt a light touch. But when her robot did the same to their palm, they grew ticklish.[8] Blakemore hypothesized that the motor command used to move the paintbrush was also sent to brain regions evaluating the sense of touch in the palm. She then had the participants control the movement of the robot arm. If it was in synchrony with the participant's movement, then it felt like a light touch; there was no tickling. But if there was a lag between the participant's movement and the robot's, then it tickled. The longer the delay, the stronger the tickle. She realized that the timing between the creation of the efference copy and the predicted sensory consequences was critical.* To correctly interpret touch, there needs to be a highly calibrated process so that motion signals and sensing are occurring close together in time. The efference copy arrives earlier (compared to the slow responses of muscles or joint receptors). With a bit of math, it is possible to show how the brain can match them up in time by delaying the efference copy by just a little bit.[9]

Blakemore's study nicely shows that the sensing of movement and efference copy signals must be exquisitely synchronized to be combined. What about intensity? How finely tuned is sensory cancellation to the intensity of the movements themselves? To address this question, Daniel Wolpert performed an experiment that was inspired by tit for tat, a punching game among adolescent boys. If you ask two young teenagers to take turns hitting each other on the arm, the only rule being they should match their punching power to what each feels, there is an inevitable escalation . . . Each hit is slightly stronger than the last until the two friends are absolutely pounding each other. Wolpert and his lab members figured out a laboratory task akin to the game. Two participants felt forces on their

*To test if the efference copy really generates a feeling of movement, I had the opportunity to perform an experiment on myself. I had a total knee replacement without general anesthesia. Instead, the anesthesiologist injected a local anesthetic into the spinal fluid bathing my lower spinal cord. I couldn't move my legs nor feel anything at all below the waist. Wide awake, I tried with all my might to move my foot, with no effect. And yet, the effort led to a feeling of motion across the top of the foot, as if the foot was shifting under a sheet—a crude reflection of the efference copy predicting what I might feel if I was actually moving.

fingers generated by the other person. They were told to pay attention to the amount of force they received and to use an identical amount of force in return. As they took turns, the forces the two participants applied increased in a steady tit-for-tat escalation. That escalation reveals that something is not quite right with sensory cancellation. Escalation comes about because the efference copy cancels out a little too much of the sensory consequences of making the hit, and so self-generated forces always feel a little bit weaker than the forces generated by someone else. Canceling out a little too much sensation is probably a good thing. It provides a margin of safety, ensuring that a person thinks she is performing the movement and not someone or something else.

The trout knew exactly how to use its efference copy to keep track of what the current was doing to its motion and, with that, how to expand its repertoire of possible actions to foil me. Out of the corner of my eye it reappeared slowly, working its way upstream with occasional taught muscular flicks. I stood motionless, spear at the ready. This time I loosened up, staring in the general area of the fish, not trying to guess where it would go. I held my gaze steady, let myself go, and lunged. As the fish's path swerved, I reflexively adjusted my throw as well. The spear found its mark and I quickly flicked the fish onto the bank. It was a much-needed meal. My change in strategy, one where I just lunged and reacted, allowed me to make a speedy correction. The hidden hand—aka the efference copy—took center stage, providing a signal that solves the second great problem of movement: how to make a speedy correction, one that seems faster than thought itself.

The on-the-fly adjustment of an action, appearing throughout a movement, is something almost all athletes rely on to catch or hit a moving target. For example, in May 2012 the Australian Samuel Groth served up a tennis ball against Uladzimir Ignatik of Belarus that was recorded at 163 miles per hour.[10] Ignatik started his return swing at a ball that traveled only 70 feet before the bounce. What if Ignatik noticed that the oncoming ball had spin on it? Then he would also need to make swing adjustments. Once the ball left Groth's racket, Ignatik had only approximately 292 milliseconds to make an adjustment before the bounce. Given the time it takes the visual system to detect the spin (about 70 milliseconds), the delays of the proprioceptive system in sensing body position (between 40 and 60 milliseconds), and the inertia of the muscles in responding to a motor command (roughly 50 milliseconds), he should have time to react to the curved trajectory in only the earliest moments after the serve. Furthermore, any conscious decision, an evaluation of whether to lunge left or right, would have cost him at least an additional 200 milliseconds; he would simply have run out of time and whiffed his return. He had to react without "deciding" what to do. Even if he avoided this decision-making step, he still faced a second problem. If his brain depended entirely on his proprioception and sixth sense to keep track of where his arm was during the swing, he would have only a late and brief moment to react to the curve of the incoming ball. To buy himself more time, Ignatik's brain could leverage his efference copy, a signal that was already providing a description of his own unfolding arm swing, allowing him to get a jump on the proprioception part of the calculation.[11] The internal feedback from his motor command was already predicting something about what his body would be doing in the very near future. Ignatik went on to win the match over Groth, who remained ranked at only 340 worldwide that year.

Ignatik's success shows how fast actions can be adjusted rapidly. In the lab, many experiments have tried to set a world record demonstrating the earliest possible moment a person can make a motor adjustment. The record may be from a study reported by J. F. van Sonderen of Utrecht University.[12] His participants pointed to a target that jumped in spatial position when the hand was beginning to move or already moving. A moving hand deflected toward the new location within 30 milliseconds! This is superfast, as fast as it takes your eyes to jump from one word to the next while reading this or three times faster than an eye blink. Other studies with similar designs have also reported superfast corrections appearing around 45 to 60 milliseconds after target jump.[13] Van Sonderen's participants had it easy because he eliminated the possibility of a decision-making step by appealing to Prablanc's trick: switching the target when the eyes were moving, when the participants were blind. If they had been conscious of the switch, they might have tried to reprogram the entire movement, essentially starting from scratch, costing them about 200 milliseconds.[14] This is one of the many mistakes I had been making when I started trying to spear the fish. I was watching and trying to change how I would throw the spear based on a conscious evaluation about what the creature might do. By the time I deliberated, it was simply too late and the fish was gone. This conflict between what the motor cortex is trying to do as it deploys the efference copy during an action and what mental deliberation contributes as it second-guesses what the goal should be helps to explain why the latter can be so subversive in sports, the pause that undermines all quick action. Athletes who play fast, dynamic sports need to learn to execute their actions "lights out," without decision-making to slow their quick adjustments.

The special contribution of the efference copy to on-the-fly corrections assumes that information about body position, coming from the sixth sense, doesn't arrive fast enough to the brain to be useful near the start of the movement. But wait.

What if the peripheral senses for position are faster than we thought, providing a quick early signal? To eliminate this nagging doubt, experimenters sought out patients with a profound loss of sensory nerves, much like Ian Waterman. One was a forty-one-year-old radiologist with an autoimmune disease that destroyed her sensory nerves. Her brain was normal. Her sensory organs were normal. But the large insulated nerves that rapidly transmit the senses of touch and position to the spinal cord were severely impaired. Thus, while she could feel a light touch on her skin, if blindfolded she couldn't tell when the investigator moved her limbs. Remarkably, she continued in her medical career. The only time she ran into trouble was if the lights were unexpectedly turned off in the radiology department: then she sometimes fell out of her chair.

To examine if the efference copy helped her make speedy corrections to her movements, she was asked to point to a target on a table in front of her. It was set up so she couldn't see her own hand while she pointed. She had to point by feel. Despite that, she was nonetheless pretty good at aiming her arm movements by dead reckoning, although she overshot the target a little bit.* The critical test was to see what happened when the target jumped. If the motor system needs peripheral sensory evidence to make a quick adjustment, then she should have a big problem making an on-time and accurate adjustment toward the new location. But she didn't. She made gorgeous, precise, and on-time corrections on par with what any of us would do.

*Without some sixth sense, her motor command was making a reasonable guess of how far to go that couldn't fully adjust for the mass and inertia of her moving arm. A little bit of body sensing would have taken care of this minor problem.

This revelation—that the efference copy is available throughout a movement—put to bed a long-standing but completely incorrect two-stage psychological model of how movements are organized.[15] Old-fashioned models (without efference copy) proposed that actions like pointing at a target start with a ballistic movement: the motor command fires off a signal to the arm, launching the hand toward a target like a ball shot from a cannon. In this model, the hand is unable to change course until sensory information, particularly vision of the hand, becomes available. This happens in the last moments, guiding the limb home. All of the behavioral experiments on efference copy and early corrections show this is simply wrong. Instead, what we find is a beautiful use of the hidden hand as an insider source of what the intended movement is, information that can adjust trajectories throughout most of an action. And the adjustments can occur quite early in the movement.

With the integration of the efference copy into movement, evolution found a way to overcome the imprecision of feedback control based on the relatively slow sensors and delays inherent in all biological systems. Watch the fast-swinging movements of arboreal monkeys or the high-speed chase of a big cat and one notices that the capacity is ubiquitous. Engineers eventually figured out additional computational solutions in the 1970s that can further enhance fast movements, calling them "feedforward control." The idea is simple. If you know something is going to happen, such as a gust of wind, leaning toward the direction the wind will come from before it hits can help to keep it from knocking you down. More formally, external information can be used to modify the behavior of a system without regard to any feedback or knowledge of the state of the system. As the fish moved, I needed to add a little English to my throwing arm to adjust the direction of the spear. For my throw, it required knowing how far along the swing already was, knowledge provided by the efference copy. But the movement also required my knowing how much English to add at any point in the swing. That is, the brain has to have a super-high-resolution hidden or internal model of where it is, what it is doing, and how much it will take to make an effective adjustment at any given time. The patients with supernumerary limb syndrome give us a clear sense of the high fidelity of this internal model. We really can generate an incredibly rich sense of where we are and what we are doing. The problem with feedforward control by itself is that the internal model has to be almost perfect to keep the throw decent or the rocket from crashing. Very quickly, the engineering community learned to make devices that used both feedforward control to anticipate changes and feedback control to keep things from overcorrecting.

After a delightful breakfast of hot oatmeal and fried fish, it was time to head off. The skies were still threatening, and I was anxious to make some miles before I might find myself tent-bound in what was looking like a coming storm. My travels that day were to a remote cirque, a ring of ridges and exceptionally high peaks. To get to them I had to ascend yet another talus field. Well over a mile long and very steep, the field was nothing more than piles upon piles of jagged rocks. This field had formed from a massive landslide in relatively recent geologic time and was notorious for its looseness. That, combined with its steepness, meant that about one in every four rocks was loose or tippy. The trick to moving across was to keep my momentum up so that if a rock started to roll, I was already well on my way to the next. I had to use feedforward control to steer my foot to the next best target whenever a rock started to move and feedback control to maintain upright on whatever unstable rock I found myself. On top of finding the best rocks to hop to, I had to keep track of where I wanted to get to on the other side of the talus field. As a goal, I picked a scraggly pine, growing almost sideways after a life in the wind.

It should be possible to find in the brain not only the circuits that create the hidden hand; there should also be places representing a desired goal that the entire motor system, including feedback and feedforward control, is striving for. In 1996, Michel Desmurget arrived at my lab at Emory University and set out to find where the brain relates a desired goal with a simple action, where the efference copy and goals meet up. He used a table with a row of lights that a healthy participant could point to. He set up a bunch of trick mirrors and lighting so that his participants couldn't see their own hand, but no matter where they reached, the target light was always visible. He switched the location of the target so the participants had to make hurried adjustments while reaching. He also used Prablanc's trick of switching targets only when participants moved their eyes, so they could not notice what he was doing. During the test, he used a homemade transcranial magnetic stimulator to apply a single powerful magnetic pulse over the participants' superior parietal lobule, a part of the brain that merges all kinds of sensory information and motor information. He applied it just as participants began moving their hand toward a target.[16] When there was no target switch, briefly turning off the parietal lobe didn't do much. However, when there was a target switch and the parietal lobe was shut down, the participants amazingly acted like automatons. They reached for the original target and made no correction until the pulse wore off, after about two hundred milliseconds. What was he to make of this? What he disrupted was their ability to link the new target location with ongoing motor behavior. Without the parietal lobe, there can be no fast, online updating. No hitting a curveball. No spearing a moving fish.

If the motor cortex is creating the hidden hand—*i.e.,* the efference copy—then there should be ways for this signal to move not only down to the spinal cord and out to the muscles but also to areas like the superior parietal lobule, where Desmurget suspected goals and command signals are related to each other. In line with this, anatomic studies show how the primary motor cortex makes extensive connections to the adjacent sensory cortex and superior parietal lobe. The input from the motor areas should arrive at these sensory areas before movement even starts. This can be shown experimentally by measuring the earliest moment when activity in neurons in the primary sensory cortex of a monkey develops with movement.[17] As one would expect, when the experimenter moves the monkey's hand, sensory neurons begin to fire as they detect movement sensed by the joints or muscles. The key observation was that when the monkey made the same movement, some of these sensory neurons would begin to fire before the arm even began to move, likely driven by a hidden-hand signal from the motor cortex arriving earlier than activity of joints and muscles.

The ability to track all these different sources of evidence, coming from the motor cortex, joints, muscles, balance organs, and vision and further to relate all this evidence with a desired goal over time is something engineers refer to as "state estimation." What is the state of your body compared to your intended goal at any given moment? A loss of state estimation renders a person adrift among her actions. At the UCL Queen Square Institute of Neurology in London, a woman was studied who had such a problem. She had fallen and been knocked unconscious from a blow to the left side of her head.[18] She developed a fluid-filled cyst in her left superior parietal lobule that slowly expanded and created new problems for her. Three years after the fall, when she reached the age of forty-eight, she sought help because her right arm and leg would drift. She said they would "fade" and become lost unless she kept an eye on them. When riding the bus, other passengers would trip over her right foot, which, unbeknownst to her, had migrated to the middle of the aisle. She was tested with a device that measured the grip strength of her hand. When she was given feedback on a computer monitor telling her the amount of force she was producing, she could track a changing goal quite accurately. She could hold a constant force accurately as long as the screen was on. As the monitor was turned off, her grip would slowly fade. Thus she had normal motor strength, a normal sixth sense of position and motion, and normal vision. But goal, movement, and sense of movement were disconnected from one another, leaving her with a lost estimate of her physical state in relation to an action.

That the parietal lobe monitors the evolving state of the body in relation to desired goals can be observed in neurophysiological recordings of parietal lobe neurons in monkeys as they track slowly moving targets with their arm. A given neuron in the parietal lobe could potentially track either the position of the arm, the direction the arm is going, or the direction of the desired goal. Furthermore, it could be lagging behind, making a prediction and leading, or be in real time with zero lead or lag.[19] To sort these possibilities out, obstacles were put in the way of the direct path to the target. Thus the monkeys had to make movements in one direction for a motor goal located in a different direction. The results were striking. Parietal neurons demonstrated a real-time coding for the desired direction relative to the goal, the here and now of where the body was in relationship to the goal. This is the essence of state estimation: making sense of all the inputs and desired goal in the here and now.

Just as state estimation is at the heart of computations that allow Elon Musk's team to land a spent booster rocket onto the deck of a rolling ship, it is essential for almost every movement we take as we land our feet on the ground. Furthermore, without state estimation, there is no way to manage more complicated actions that require the person to remember where and when to move all of the body, not just the parts that are in view. As I moved through the talus field, I often had to hop over taller rocks to get to a good foothold. Visual feedback and state estimation helped me to make sure I lifted my lead foot to a sufficient height. But what about my trailing foot? I couldn't see it—it was behind and underneath me—and I was too busy looking ahead. Without ever having to think about it, my brain remembered the obstacle and lifted my trailing foot to an appropriate height so I wouldn't catch my toe and fall. This is the epitome of state estimation in action. And with that, I could safely bound across the rocks, quickly reaching a safe haven on the far side.

5

Pulling Strings

In the morning, bright and early, Pinocchio started for school. Imagine what the boys said when they saw a Marionette enter the classroom! They laughed until they cried. Everyone played tricks on him. One pulled his hat off, another tugged at his coat, a third tried to paint a mustache under his nose. One even attempted to tie strings to his feet and his hands to make him dance.

—THE ADVENTURES OF PINOCCHIO[1]

IT HAD SNOWED through the afternoon—each snowflake slowly melting into my clothes until I was a soggy rag. A chill was quickly setting in, it was time to stop. Fifty or fifty thousand years ago, someone in this situation wouldn't have thought twice about lighting up a nice hot campfire to dry out. But trees grow slowly at high altitudes, and the eventual scarcity of wood culminated in the outlawing of campfires above ten thousand feet of elevation. Unlike our ancestors, the mountain traveler now relies on a camp stove, and when it is brought inside, a tent can become a sauna. I had a general-purpose stove, designed to burn just about anything flammable: kerosene, jet fuel, leaded gas, or alcohol. As I ignited it, the flame sputtered, producing a pathetic cool yellow glow that billowed black fumes smelling like an old diesel Mercedes. The diagnosis was easy. The fuel line was loaded with carbonized crud that prevented the fuel from vaporizing. After shutting the stove off and cooling it down I pulled it all apart. Lots of little O-rings, gaskets, screws to hold bottom plates on, tubes, cleaning wires, and nozzles were arranged across my sleeping pad, a deconstructionist art exhibit in miniature. I cleaned out all the tarry gunk and started the reassembly. It went well until one of the final steps. I had to hold a screw, a nozzle assembly, the base, and tubing all with one hand while the other hand managed a knife that was substituting for a screwdriver. This would have been easy with three hands. With two, my project became a tottering jigsaw puzzle of instability. I had to push one set of parts with my ring finger and pinky against a second bit stabilized by the middle finger, while the index finger and thumb held the screw. This maneuver required much more than my fingers were familiar with. Each time I made an attempt to insert the screw, one of the three parts would fall to the ground. It was getting dark and I had to use a headlight to see what I was doing. A little panic swept over me. What if I couldn't put the stove back together? Cold, wet, and no way to cook. Sure, I would survive this discomfort, but it would be unpleasant. I kept at it. I broke the necessary finger actions down and practiced, trying to get my fingers to line up at the same time so as to complete the task at hand.

In the course of the frustrating effort, I was teaching my hand to isolate extremely awkward movements, little adjustments needed to form an entirely novel shape. In part, the new shape was built from familiar ones, like holding a screw between two fingers rather than a finger and the thumb. But it also required unique movements. It really wasn't all that different from learning how to knit, sew, or crochet—each a discipline of quirky little finger movements. Somehow, my brain was able to search through an infinite sea of possibilities to locate the few that might work. This wasn't blind luck. There was something that was organizing all my actions so that the search would be easier, so I could learn a new skill in minutes rather than after years of trial and error. The organizing principle was playing off two very different ways that physical skills are learned. On the one hand, there is a fast-track way: a reassignment of known movements to make new ones. On the other hand, there is the learning of hitherto unknown movements in order to achieve an unpracticed action. This tension between familiarity and novelty is reflected in the organization of the motor system across many levels, from the design of the muscles and skeleton, to the wiring from brain to muscle, and, of course, to the patterning of brain activity splayed across the surface of the cortex.

There are 206 bones in the human body, linked by 360 joints. Imagine you are a puppeteer and you are trying to control a marionette made out of so many parts. How would you attach the strings to make your marionette dance or wave hello? What tricks might you use to make it easier to control? This is the heart of motor control: figuring out the organization of the strings and the role of the puppeteer. One trick that almost every puppeteer uses is to tie some of the strings to one another so there are fewer of them to pull at once. If you tie them in just the right places, then the limb can make a single coordinated movement like a step, kick, or reach. This same trick is often used in nature, albeit with nerves rather than strings. As an example, watch the cephalopod *Wunderpus photogenicus* moving about. It is a particularly glamorous octopus with very sophisticated behaviors, including hunting and competing for survival in an ecological niche that has some ruthless vertebrates.[2] *Wunderpus* has no fixed joints in any of its eight arms, giving it even more ways to bend its appendages than we vertebrates ever could.[3] Despite the lack of rigid bones or joints, it is able to reach, hold prey, and propel itself perfectly well. How does it do it? It could have relied on sophisticated algorithms in the brain and nerves to meticulously engage each muscle, like a puppeteer carefully pulling countless strings of every individual segment. In theory a cephalopod could have been informed by this very sophisticated kind of control. It has a pretty big brain, with roughly three times more neurons than a dog has. But it uses most of those neurons for other things, like vision. It has drastically fewer neural connections from the brain to each of the arm muscles, sharply limiting any tight control over the individual segments of the arm. Instead, it is endowed with an orderly wiring diagram of nerves built into each arm that cause the muscles to fire in particular patterns, leading to highly consistent behaviors.* The octopus overcomes the complexity that stems from a lack of joints and bones by strategically using only a few nerves, akin to the puppeteer tying most of the strings of the marionette together to create only a few stereotypical movements. The trade-off is that, just like the simplified marionette, the octopus possesses a very narrow repertoire. If it catches prey on a suction cup out at the end of the arm, the octopus brain can trigger the nerve plexus of that particular arm to make just one kind of movement that sweeps the prey to its mouth. If the prey is stuck on a suction cup in the middle of the arm there is another movement.[4] Critically, there are no in-between movements, just those for two targets: prey at the end or prey in the middle of the arm. It has to

*This is a wonderful example of embodied cognition. Evolution constantly uses an engineering strategy of sticking some of the problems of control out in the body parts rather than trying to solve everything in the brain. Embodied cognition was first recognized by Rodney Brooks when he was designing artificial intelligence systems for robots. He realized that it can be better to design some of the problem-solving out in the periphery rather than inside a central processor. This solution is now widely adopted in many man-made systems, from the anti-skid brakes in a car to the flight control systems inside a commercial airliner.

make do with just the two movements. All eight arms act pretty much the same way. There is no improvisation, no creativity, no flexibility.* Nevertheless, we see from *Wunderpus* that with just a few simple commands from the brain, all the actions necessary for its survival are fully enabled in a wonderfully beautiful being.

Cephalopods aren't the only creatures that can act without much input from a brain. If you put a small piece of meat in a cup, a cat lacking a motor cortex can still reach in with its paw and scoop it out.[5] Admittedly, the grasp of this unfortunate cat is not particularly refined. The paw and claws form a stiff rake without much movement or shaping of the paw. Amazingly, cats without a cortex can also walk on a treadmill, albeit with a fairly stereotypical walk that is adjusted based only on what they feel through their feet. They won't anticipate changes in the surface or step over obstructions. Thus, even in mammals, basic programs like reaching, grasping, and walking can be fully generated from a combination of hardwired brain-stem and spinal-cord signals, much like the reflexes in the octopus.[6] But what about us? To what degree do people draw on these hardwired solutions?

A newborn baby is a little bit like the *Wunderpus.* A baby can reach out toward bright shiny objects and will automatically open his fingers.[7] Look closely and you can see an expected motor pattern at work, as with the octopus. You will also notice other simple behaviors, like sucking when you tickle the baby's lips or opening of the mouth in anticipation of an arriving thumb.[8] At birth we are all endowed with some rudimentary

*There are countless online videos of street entertainers with marionettes that can dance quite well. How the strings are tied determines whether they dance hip-hop, ballet, or disco. The anthropomorphic qualities are enticing for a short while until one realizes the puppets are limited to only a few dance moves.

and automatic motor programs, already hardwired into the nervous system. These primitive behaviors are probably controlled by the spinal cord and brain stem, with minimal reliance on the motor cortex or other higher brain areas. As the motor areas in the brain develop, a baby learns an ever-growing number of more sophisticated motor behaviors. By sixteen weeks it shows off precise reaching movements and coordinated grasping, and by two years, its primordial movements are largely hidden. By adulthood these primordial actions are almost impossible to elicit. You can unmask a little bit of this wiring by learning to belly dance. The novice belly dancer will learn to control her abdominal muscles in an alternating diagonal pattern that is nearly identical in form and frequency to what a salamander does when it runs.[9] With more training, an expert dancer can generate top-to-bottom rhythms in a form and frequency that are the same as the swimming motion of salamanders and lampreys. Deep inside, we are endowed with wiring diagrams and controllers that were created far in the evolutionary past, stretching all the way back to the appearance of the most basic forms of locomotion in vertebrates. As the brain learns, these become hidden as other controllers take hold, but they are there nevertheless.

In the neurology ward, I sometimes get glimpses of other primordial movements, particularly in patients who have significant frontal lobe degeneration or injury. They might show grasping reflexes, or sucking reflexes with the lips, confirming that these infantile behaviors never really go away but are just hidden by more dexterous skills. People who have suffered a stroke and are weak will also develop contractions of the arm or leg into rigid postures. These too are basic programs generated in the brain stem and spinal cord designed to stiffen the body during movement and that run unchecked after stroke. There was hope that such patients might be able to leverage some of these primordial programs as a way to overcome their loss of dexterous movement. Unfortunately, decades of intensive rehabilitation experiments show that for humans, these deeply situated programs are hard to recruit and they can't do quite enough to compensate for a damaged cortex. We humans need a motor cortex to pull the strings for most of our goal-oriented actions.

To go beyond primitive reflexes, the baby needs to acquire control over individual muscles, to pull the strings in ever more complicated patterns. In people, the connections are a little bit more complicated than a marionette's strings. People need neurons to activate muscles that move the bones, roughly seven hundred muscles in all. Unlike the single string connecting the puppeteer to his marionette, the neurons don't go directly from the brain to the muscle. Instead, the most direct connection requires two neurons linked in a chain. The first goes from the brain to the spinal cord and the second goes from the spinal cord to the muscles. Given that, it helps to know if the connections are one to one, with each cortical input talking to one motor neuron in the spinal cord, which then goes to one muscle. It turns out it is wildly disproportionate. We know from anatomic studies that there is a massive number of neurons in the cortex and brain stem that talk to the spinal-cord motor neurons. Furthermore, it might be the case that some parts of the body that we think are better at making fancy movements (like the hand) get more nerves to the muscles than what look to be dumber body parts (like the upper arm).

Only recently, a team of anatomists was able to get an accurate estimate of how many nerves go into the arm from the spinal cord. To figure this out, they had to examine fresh human nerves using special stains. They carefully dissected out all the nerves passing into the arm in patients who had just died and whose heart or liver was going to be removed for organ donation.[10] This is a noble kind of science that is rarely done any longer: painstaking, difficult anatomy. The researchers discovered that the nerves passing into the arm carry a total of 350,000 axons (the individual "wires" that carry information for each neuron). Of those, only 35,000 carry motor signals from the spinal cord to the arm muscles. All the rest are sensory neurons, allowing a person to feel position, touch, temperature, and pain. Of the 35,000 axons carrying motor signals, most go to the shoulder and forearm; only 1,700 make it into the hand.* Thirty-five thousand is in fact a surprisingly small number for such phenomenal limb dexterity. It shows that the number of motor neurons passing from the spinal cord to the muscles is determined mainly by the size of the muscles, not by any particular functional requirement like hand dexterity. From a practical standpoint, this means that the new skill I was trying to learn to fix my stove wasn't likely to be coming from adjustments to those few neurons passing from my spinal cord into my forearms or hand muscles. The skill must have been coming from higher up, in the way all those nerves descending from the brain and talking to the spinal cord were commanding this very limited complex of connections to the hand muscles. If those 35,000 axons passing into the arm were marionette strings, each string would have one hundred or more different puppeteers trying to control it.

Given the massive number of neurons that project from the motor cortex to the spinal cord, what kinds of instructions do they send? Any one of them might branch and send commands to many other motor neurons in order to make different muscles

*This helps to explain why hand function is so easily lost in carpal tunnel syndrome. Since there are so few axons entering the hand, even a mild compression of the median nerve can create big problems.

move together. Or they might influence only one muscle, a far simpler proposition. This is a straightforward question to ask but a really hard one to answer experimentally.[11] When the motor cortex is recorded, it becomes clear that most of the neurons aren't even sending signals to the spinal cord. Most signals go to other parts of the brain. The challenge is to find just those neurons that take the most direct path to the spinal cord to drive the motor neurons that move the muscles. To do that, an experimenter must also record electrical activity in the muscles and find those neurons and muscles that start firing at almost the identical time. He is like the puppeteer who has to pull one string at a time to see which segment of the marionette twitches first. This is painstaking work, and it can take years to distinguish these neurons from all the other neurons. The results are beautiful: nearly every one of these command neurons with the strongest influence over the muscles drives only one muscle. Each cortical motor neuron will influence only one muscle. This is good news in that it simplifies the wiring diagram. It also provides enormous freedom for the motor cortex to create any sort of movement it wants (provided the bones and joints will allow it!) and to individuate, to isolate the contractions of particular muscles in new ways. This is what I was trying to do with my hands to fix the stove. It is also what determines the capacity of modern dancers to create surprisingly novel motion and allows for the spectacular control of gymnasts, musicians, and surgeons.

How should the chorus of puppeteers work together to decide which strings to pull and when? For decades, there has been this notion that the neurons in the cortex are organized around the direction of movement: one team of neurons would make the body move in one direction, and another team would move the body in another direction.[12] If this is the only way the brain is organized, then a neuron should be silent for movements in directions that it doesn't prefer. However, when a group led by Peter Strick at the University of Pittsburgh looked at this issue carefully, they found something very different.[13] As the older literature had suggested, each cortical neuron would fire more for a particular direction. For example, it might fire most when the monkey flexed its wrist. But it wasn't necessarily inactive with respect to all other directions. It might also fire when the hand was moving in the exact opposite direction! Furthermore, the overall pattern of firing would shift as the monkey changed arm posture. The experiment showed that neurons are doing far more than just moving the body in one direction or another. A given neuron might also be involved in other aspects of an action, like stiffening the limb. Stiffening is critical. For almost any action you want to do, there is always a bit of stiffening occurring somewhere in the body to stabilize the movement. Hold a bowl of cereal in front of you. You will stiffen your arm so that if you are jostled by someone you won't spill any milk. In other words, while it is still possible for the cortical neurons to form teams to make a limb move in a particular direction, they are also ensuring that other aspects of an action like stiffening are happening as well.

How complicated! Each cortical motor neuron can be involved in many kinds of actions, not just the direction of movement. And more: it turns out that any given action can be made by timing the string pulls in an infinite variety of ways. Imagine sticking your arm out to hitchhike. You could lift your arm, straighten your elbow, and then stick out your thumb. Or you could straighten your elbow, stick out your thumb, and then lift your arm. The combinations all leading to the end result of the hitchhike gesture are endless. This was first recognized in the 1960s by the Russian physiologist Nikolai

Bernstein, and he called it the "motor equivalence problem."[14]* It quickly became apparent, particularly for those trying to write software that could control the movement of humanoid robots, that there could be no unique computational method to solve the equivalency problem. The implication was that there was no way the brain could ever perform the calculation to generate a unique set of commands to move the muscles in a desired action without adding some additional rules or guidelines. Insofar as many of us move around perfectly well, there must be some additional computational tricks used by the brain. Researchers spent decades searching for mathematical rules that would make this hard computational problem tractable and simple.[15] In retrospect, it is clear that the search for a solitary mathematical algorithm to program the vast repertoire of natural movement was probably misguided. Without knowing any math, Mother Nature solved the equivalency problem by making a bunch of design choices for the motor system—in the way the body is built, in the wiring diagrams, in making software compromises, and in the kinds of movements that are possible—all resulting in a super-specialized system that, by design, drastically simplified the equivalency problem. These hard choices of evolution are only barely beginning to be understood.

One way that natural selection can deal with the complexity of motor control is by making a system that can generate only the actions that are needed to survive in a particular environment. *Wunderpus* needs only a few definitive movements to get by, and it doesn't make use of any others. Even though human hands seem exceedingly complex, it's not necessarily

* His definition has largely been replaced by an equally opaque terminology: the "degrees of freedom problem."

the case that we need to use our hands in an infinite variety of ways. Maybe a few hand configurations might do the trick for most actions, and in that case, the brain can focus mainly on refining control. If so, then most of what we do with our hands would presumably be determined by the kinds of objects that we manipulate in the natural world. Long before there was high-tech motion-capture technology, Noriko Kamakura of the Tokyo Metropolitan Institute of Gerontology found an elegant way to investigate human hand shapes, reminiscent of Gilles de la Tourette's inked footprints for categorizing gaits.[16] Kamakura dipped ninety-eight common household objects like scissors, knives, plates, and chopsticks in a bucket filled with a mix of glue and black ink. Her participants picked up the black objects as if they were going to use them, then put them back down. She took photos of the sticky ink patterns that each object imprinted on their fingers and palms. She then made a qualitative comparison of all the different ink patterns and recognized that the ninety-eight patterns could readily be boiled down to fourteen prototypic grasps, falling into four overall groups: power grips, precision grips, scissor grips, and disc lifts. In other words, of the infinite ways you might use your hands around the house, you typically need only fourteen grips to handle most objects. For the brain, that means there is a far shorter list of hand actions it needs to learn how to generate routinely. Twenty years later, Kamakura's fundamental discovery, based on the simplest kind of experimental observation, was replicated with far more sophisticated motion-capture technology.[17]

The man-made objects that Kamakura used were specifically designed to be easy to handle. We don't know if they are the kinds of objects that influenced the evolutionary design of the hand in the first place. It's pretty clear that early humans weren't using scissors, and there are only so many natural objects that they could have manipulated. For comparison's sake, we might consider other primates. As an undergraduate at Princeton, Nicholas Macfarlane had the opportunity to study what monkeys in the wild do with their hands when left to their own devices.[18] Some Old World monkeys like the rhesus macaques are endowed with a relatively long thumb. Indeed, the proportion of their thumb length to finger length is quite humanlike. Do they use their hands the way that we humans do? Macfarlane examined more than fifteen hours of film of 157 macaque monkeys living in the semi-wild on Cayo Santiago, a tiny island off the coast of Puerto Rico. In what must have been an exhausting analysis, he reviewed every frame of film and found all the events where monkeys used their hands. He was able to identify 5,554 clear-cut instances. Notably, the monkeys were roaming around making natural movements with objects from the island like food and rocks. They were not handling man-made tools or objects in a laboratory setting. Macfarlane lumped the different arm-grasping actions into functional groups, based on the contact points of the fingers and palm with objects, reminiscent of Noriko Kamakura's approach. Not surprisingly, Macfarlane caught the monkeys doing a lot of climbing in trees, and to do this they used only a few types of power grasps akin to the way we squeeze a pull-up bar. Most of the power in this kind of grip comes from the four fingers wrapped around the branch rather than from the thumb. More surprising was the variety of grasps the monkeys used for manipulating objects. Compared to humans, monkeys are handicapped because they lack a styloid process in their metacarpal bones as well as a few of the hand muscles that humans have. This limits the power of many of their hand grasps. Nevertheless, they routinely use their hands to manipulate objects in a multiplicity of grasps. Macfarlane was able to distinguish eleven distinct types, which he gave self-explanatory labels such as "pad-to-pad," "two-hand roll," "between-fingers," and "three-tip." The monkeys demonstrated amazing ease in switching between different types of grasps, underscoring their sophisticated behavioral flexibility in accomplishing action goals. A direct comparison between Kamakura's pictures of human hands and Macfarlane's monkey hand grasps is quite informative. Eight grips were identical. The other three were almost identical. Whenever they were using variations of a power or precision grip, the two species were effectively using the same hand shape for a large fraction of hand actions. Monkeys didn't use scissor or disc-lift grasps. There were no grasps made by monkeys that people can't do. Old World monkeys and humans share a repertoire of hand functions that overlap; both species share a similar basic set of prototypic grasps. The problem of reducing the number of possible functional hand shapes from infinity to a small manageable set has been working itself out for a very long time in primate evolution. Natural selection found a remarkable simplification with a particular hand design, dodging much computational complexity. Within the motor cortex, there needs to be only a baker's dozen or so teams of neurons in order to create most of our natural hand movements.

This may be true for the rest of the body. We rely on a fairly small number of movement patterns for walking, running, climbing, and using our arms, easing the challenge to the brain to act as puppeteer. But even as the number of possible actions is reduced, the motor equivalency problem remains. Frog legs are a particularly good way to study how a creature can overcome this problem. A frog doesn't have a lot of muscles, and it engages in only a few actions involving its leg—jumping, scratching, or swimming—making it relatively easy to study. These are nicely stereotypical behaviors, and so there should be a very reproducible pattern of muscle activations for these three actions rather than an infinite variety of firing patterns. To see if this is true, Emilio Bizzi and his students at MIT used tiny electrodes to record activity from all the muscles in the frog leg as it made these different kinds of movements.[19] For each action, they found a very orderly, reproducible pattern, with each muscle firing at particular times. And unlike Bernstein's worry about motor equivalency and an infinite variety of possible ways to time the firing patterns, for a given pattern the frog used the same pattern every time. And unlike the octopus, where the nerves were hardwired to fire for one action or another, in the frog each nerve could be used in a variety of actions.

Bizzi's team used a data-analysis method that reduced all the individual muscle sequences for the different actions into groups. These simplified patterns form a "basis set." From this basis set it is possible to re-create all of the original and more complicated patterns of muscle activity that lead to full-blown actions. Bizzi referred to each member of the basis set as a "muscle synergy." The synergy reflects fixed patterns describing the order in which the muscles get recruited over time. Just as a single muscle doesn't typically make an interesting movement of the limb, a single synergy is still not enough to generate a complex action. By combining a bunch of overlapping synergies in a particular order, it is possible to reproduce the frog's swimming, jumping, and scratching movements. In doing so, the master recipe for making the frog leg kick requires only four or five instructions. Bizzi made a clear case that in the frog these synergies are not a result of the way the muscles and bones are connected, or just a primitive reflex. Instead, they come about as a sophisticated design principle for simplifying complex control of the muscles.

But are muscle synergies observable in complex actions like grasping? We know from Kamakura that there is a manageable number of hand shapes for most actions. Are they associated with inherent synergies as well? To test this, Bizzi's team wired nineteen muscles in the arm and hand of a monkey that his team had taught to reach out and grasp objects of various sizes and shapes.[20] As with frogs, the team could boil all of the variation in muscle activity down to a small set of synergies. Remarkably, using just three synergies, they could re-create more than 80 percent of the individual muscle activity, irrespective of the object. Since then, analogous patterns have been discovered in humans.[21]

One of the appealing ideas that emerges from the theory of synergies is that it provides a way for anyone to quickly build new movements from old ones. At the same time, it is possible the brain is actually making entirely new synergies to use as building blocks. These possibilities are hard to detect in actions with the hands, which are quite good at making novel individuated movements. A better way to see how synergies appear or are combined with learning is with a task involving most of the body. Andrew Sawers had the great idea of measuring activity of sixteen leg and trunk muscles in people as they learned to walk on a balance beam.[22] The clever part of his experiment entailed recruiting both ballet dancers and untrained novices. The dancers were really good, all with more than ten years of experience, and some were members of the company of the Atlanta Ballet. Not surprisingly, the dancers were better at walking along a narrow beam. The question is why. Zooming in on their muscle patterns, he found two possible explanations. First, dancers had more muscle coordination patterns—that is, more muscle synergies to work with compared to novices. Second, they were more adept at using each of their different synergies to manage this new physical challenge. The results raise the question, How much dancing does it take to build these added synergies?

In another study, Sawers had a small group of patients with Parkinson's disease take tango lessons for three weeks. As with many other emerging studies, the dance lessons for these patients were particularly effective at improving balance and gait performance.[23] Again, the question is why. In this case, after only three weeks the patients showed no evidence of new muscle synergies. Rather, they were better at recruiting the ones that were already available. Although it is possible that the disease may slow the ability to make new synergies from scratch, a more likely explanation would be that it takes months to years to do so. As we age, there is a propensity to not capitalize on all of the available synergies to control even the simplest of movements, like pointing at a target[24] or rising from a sitting position.[25] It's probably wrong to think these changes occur simply because the body is old. Rather, physical inactivity with aging is likely to be a major culprit. Inactivity is recognized as a significant risk factor for changes of synergies that lead to increased slips during a walk as well as reduced ability to catch oneself during a fall.[26]

Patients with stroke and motor weakness provide crucial evidence that the motor cortex is involved in organizing many natural actions based on muscle synergies. In one study, careful recordings were taken of muscle activity in the arms of stroke patients while they performed various tasks. From this, the underlying muscle synergies were identified.[27] For those patients who had suffered a mild stroke, the kind that disconnects some of the motor cortex from the spinal cord, the weak arm displayed normal underlying motor synergies, but the way those synergies were being recruited and combined was messed up. Groups of muscles were recruited in the wrong order or with incorrect intensity. Patients with devastating strokes damaging the motor areas told another story. For them, even the motor synergies were lost. Given the involvement of the motor cortex in organizing movement as synergies, it has been argued that the cortex itself might be organized into modules that control different synergies.*

The notion of synergies really began by looking for patterns in the way muscles are engaged to perform familiar movements. At the same time, an entire generation of neuroscientists has chased after an organizing principle of the motor cortex, to determine if the spatial arrangement of neurons has a relationship to any kind of familiar movement, whether it is a thrust in a particular direction, a posture, or the twitch of an individual muscle. The latter chase really began in the 1930s with the first description of the homunculus—*i.e.,* the "little man." This is the organizational principle, proposed by the pioneering neurosurgeon Wilder Penfield, that the motor cortex is arranged in a reproducible topologic map, with the neurons going to the foot near the top of the head, then the leg, arm, hand, each of the fingers, thumb, mouth, and body all in a neatly ordered row.[28] The homunculus implied that there was a clean, one-to-one mapping between a cortical map of the body and the nerves connecting to each of the muscles. If you go back and look at Penfield's actual drawings of what he found in his different patients, however, the first thing you notice is that the map for any given patient is not so tidy. There are multiple separate places on the cortex where stimulation moves a hand, and multiple other places where stimulation moves an elbow, and so on. The second obvious thing is that there is a lot of variability from person to person. Penfield wanted to make a general conclusion of how things were organized, but he was faced with a very messy mapping problem. Thus, with some artistic license,

*Whether this modular organization actually exists in the mammalian brain remains a controversial and challenging question in neurophysiology research.

the homunculus was born: an idealization obscuring both the multiplicity and individuality of the actual cortex.

At a crude scale, he got it right; the homunculus is clearly present. Every neurosurgeon knows there is a sweet spot about three inches above the ear where stimulation will drive the hand, another closer to the ear that will drive the face, and another near the top of the head that moves the leg. Early in my career I used brain scans to map definitively these locations in healthy people as they moved their arm, face, or leg, showing the same general pattern that Penfield observed.[29] Within the upper limb it was possible to distinguish the shoulder, elbow, wrist, finger, and thumb, just as Penfield had proposed.[30] And indeed, every neurosurgeon knows and relies to a certain extent on this crude map. It remains an invaluable tool for confirming the location of the motor cortex so that the brain surgeon doesn't make matters worse by inadvertently damaging it. Neurologists routinely use this knowledge to make predictions about where a stroke might be located in a patient based on the patterns of weakness they develop. There can be loss of strength of the face, the leg, the arm, or any combination of these body segments. In many cases, the clinician can make exceptionally precise predictions about which portions of the motor cortex or its projection to the spinal cord will be damaged based on these segmental patterns.* Nevertheless, at this crude resolution, there is no evidence for any sort of modularity that could help us know how the cortex is making refined actions.

*A spectacular example of this is the "man-in-the-barrel syndrome." In this case bilateral hypoperfusion of the brain leads to hypoxic ischemic injury to bilateral border-zone cortices. This is a band of tissue damage that runs through the cortex at the level of the arms, sparing the hands, face, and legs. Thus the patients have weakness of the arms but not the hands, as if the arms were inside a barrel with the hands sticking out below.

At a slightly more highly calibrated spatial scale, Michael Graziano discovered that if he stimulated the motor cortex of a monkey more intensely than with the electrical current that neurosurgery patients receive, he could get the monkey to do something with its limb. As he stimulated a particular spot, the limb would immediately adopt a very specific posture. For example, it might move its fingers to be right in front of its mouth. Surprisingly, the monkey would adopt this same posture whether the stimulation started with the hand to the right or the left, above or below. The stimulation determined the end posture irrespective of the starting position. Moving the stimulator led to a different posture. These weren't synergies or modules. Graziano was controlling end posture alone. We still aren't quite sure how to interpret these intriguing results. Some have said that all movement is really just the control of postures, with one after another endlessly strung together. Others find the idea that posture alone is the end-all explanation of movement to be fundamentally dissatisfying. It lacks the attention to granular detail, the subtlety that one might find in the movements of a surgeon or a seamstress.

An important counterpoint to synergies is individuation, the refined control of individual muscles providing each of us with exceptional dexterity that surpasses that of all other species. There is nothing new about dexterity in human evolution. It is probably as old as hominins. At least one hundred thousand years ago, our species was delicately carving shells into gorgeous beaded jewelry.[31] This level of detailed fabrication requires a precision that is far beyond the types of actions that we think of when we talk about muscle synergies. Individuated movements require a different design to the motor cortex than modules. By using progressively smaller electrical pads to stimulate the cortex of patients undergoing brain surgery, this alternate began to emerge.[32] It was found that within the broad hand/arm region of the motor cortex in any one patient, there were many locations that would cause a hand or arm movement when they were stimulated. And they were all intermingled, harking back to Penfield's original observations. The same kind of fractured homunculus, with many little arms or hands mixed about, was found in monkeys[33] and in people.[34]

This intermingled spatial arrangement of neurons is fundamentally different from the modular view. It is far more disorganized as a spatial map. Nevertheless, the mixing of different body parts could be a great way for the brain to control all sorts of movements across different joints at once. It is also a design principle that helps to explain how we might learn movements that seem impossible at first. The intermingling allows for all sorts of new combinations in the way that local groups of neurons can fire together, leading to the creation of novel combinations of muscle contractions and entirely new actions. In this view, the motor cortex is really optimized to make all kinds of movements; whether they are built from old synergies or from entirely new patterns, there is no need for a modular organization for the cortex to do this.[35]

To allow for this flexibility, other brain areas are almost certainly needed to provide instruction to the motor cortex to organize the ad hoc teaming of neurons to generate new actions. There are plenty of premotor and parietal brain areas with input to the motor cortex that are available to do the coaching.[36] In this view, the motor cortex neurons are like a bunch of intramural athletes who can be chosen to play different sports and on different teams every afternoon. They are free agents, capable of playing many positions. The trick for this to work is that all the coaches in the other areas of the motor cortex have to know what each player can do. The free-agent idea is better suited to our intuitions about what it takes to be truly dexterous: to be able to make highly specific, minute, and customized movements.

The free-agent concept as applied to the motor cortex is borne out in part by an explosion of discoveries involving brain-computer interfaces. The most dramatic results stem from both monkey and human research in which an array of micro-electrodes is implanted into the motor cortex. The goal of early versions of these experiments was to determine if an interpretable signal could be obtained from the motor cortex in a person or monkey trying to move and to decode that signal to drive a robot arm to mimic what they wanted to do.[37] The most dexterous results have been obtained from a monkey. Just by "thinking," a monkey can reliably drive a robot arm to pick up a bit of food and bring it to its mouth to feed itself.[38] A fortuitous discovery was the realization by multiple investigators that they could get away with fairly simple computer algorithms to decode the activity in the motor cortex neurons to make the robot arm work. Rather than the computer doing the decoding, the brain with the electrodes in it did the work for it. Within minutes, the motor cortex neurons reorganized their firing patterns on their own to learn how to control the robot arm.[39] The fast learning suggested that many if not most of the underlying neurons that would normally drive the muscles of the body really are free agents, able to be retasked to move a robot arm just as readily.

As I struggled to put my stove together, what was I actually learning to do? Was I recombining old synergies in new ways, making new synergies from the bottom up, or finding new ways of controlling individual finger movements? Given that it took me only thirty minutes to succeed, I was most likely repurposing known synergies and maybe learning to isolate a few simple finger postures. This tension between exploiting what is already known and starting from scratch, whether to build synergies or to isolate particular movements, is present whenever a person tries to learn a new skill. It is a tension that every great trainer, coach, and instructor is intuitively aware of. In working with a hockey player already endowed with lots of stick-handling skills, the coach can readily repurpose his known synergies so he can create a similar skill like a golf swing. This kind of learning can be quite fast.* Once a skill is learned, the coach can work with a very advanced athlete and help him isolate specific hand or foot movements to make a more graceful or precise action. In contrast, for a novice, the coach will often break the action down, since the novice is trying something built from a pile of poorly learned or inappropriate synergies. The novice might need to learn new synergies from scratch, a process that might take years. Clearly, these different ways of shaping performance require different kinds of practice and work on very different timescales, from a few tries to a few years. While we are beginning to use brain scans to identify how different premotor and motor areas change their firing patterns as learning takes place over these very different timescales, we still lack a general theory of skill learning that adequately captures all these radically different timescales.[40] We don't know when the cortex will act like a coach aiming for quick fixes or one that breaks movements down. Nor do we

*It is common to refer to a well-learned physical skill as a "muscle memory" in recognition that it can be performed with little thought. However, the muscles are as dumb as a brick. Skill depends almost entirely on how all the premotor and parietal areas that act as coaches shape the commands coming from the motor cortex.

know why some people will work for weeks on a given skill and others quit from frustration after a few halfhearted tries. I was lucky that evening. The new skill I needed to fix my stove didn't require a year of practice. After thirty minutes, I could hold the parts and use the screwdriver at the same time. Finally, it worked! I fired up the stove and it roared with a gorgeous blue flame, as hot as any kitchen range.

6

Perspectives

A pool, a stone, a bush, a rise of ground so slight as to be hardly perceptible, objects which a European scarcely notices, are deeply impressed in their memory, and serve years afterwards to guide them over the trackless and desert steppe.

—FERDINAND VON WRANGELL[1]

THERE WAS NO sunrise. Only shadowy gloom in all directions. I was encased in a thick blanket of clouds that clung to the mountains like a London fog, enshrouding everything near and far. Yesterday's snow had turned to a night of rain, rendering everything cold and damp to the touch. It was impossible to tell where the sun was rising or any cardinal direction. I could only infer what might be a westerly direction from the orientation I had pitched my tent in the night before: the door had faced a setting sun. I was up on a high plateau that hung like a broad shelf off a now invisible peak. A few miles away, the mountains walled me off. In the opposite direction I was corralled by a long cliff edge. At least that is what my map showed. I really couldn't see more than a few feet in any direction. I needed to cut a diagonal across the shelf and find the one good spot to drop off the cliff, a tight slot providing a welcome portal that would lead down into a valley far below. Descending from the edge anywhere else would be a horror show of tenuous down-climbing on slime-covered bluffs. I didn't have a compass, but I could still draw a compass rose in the dirt by using the orientation of the tent. I studied my paper map, trying to find any topographic features that might serve as landmarks. I had to convert a simple line drawn on the map into a mental journey with me positioned at the center. There wasn't much to work with on the map and not much to see around me. Picking the most likely direction from the compass rose, I headed off, dead reckoning toward the southwest. I was engaged with a gigantic three-dimensional puzzle, ready to trick me at a moment's notice. On a clear day the hike would have been trivial. There would be points of orientation, distant mountains or passes, beacons calling me on, like a lighthouse signaling safe harbor. Normally I could create different kinds of maps in my mind that would allow me to form a coherent frame of reference. But now, in the cloudy murk, all I had was an arrow in the dirt pointing into the mist and nervous hope that I could keep myself going in that one direction.

The problem I faced was not unlike the problem that any forager faces, whether a human or an ant. The ant leaves its nest and wanders about, trying to find some food. Then it makes a direct path back to the nest. How does it calculate that straight line and, once on it, keep itself going straight? Without markers like scent, light, or geography, the animal or person can rely on pure self-guidance. This is the most egocentric form of navigation, one that is based on a frame of reference centered entirely on the self with no cues from the external world. For it to work, the ant or person would have to be able to sense the tiniest change of direction, using signals from its balance organs or subtle cues from the sixth sense to counteract the changes. But is it even possible to detect these deviations and compensate for them precisely?

Debates have raged as to whether this kind of autonomous

guidance without an underlying map of some sort is possible in living creatures. In theory, the math that describes this kind of control says that it is possible, at least under ideal conditions.[2] However, theory doesn't always capture reality. The matter was recently tested under natural conditions when the researcher Jan Souman took hikers into the dense Bienwald of Germany. He challenged them to simply walk in a perfectly straight line without the aid of a compass or map.[3] He tracked their course through the woods by GPS and found that on a cloudy day, the hikers would quickly start to zigzag. Eventually, they began to walk in circles about a half kilometer in diameter. He repeated the experiment at nighttime in the Sahara Desert. Here too, the hikers lost their sense of direction and ended up walking in circles. Thus, while theory says it is possible to autocorrect, in the real world people are profoundly bad at keeping themselves pointed in one direction if left to their own egocentric frame of reference. Look a little bit harder at the math and the reason why becomes clear: feedback control can be exquisitely sensitive to noise. For this kind of pure guidance, just a little bit of noise drastically alters the ability to detect changes of direction.[4] A walker is full of noise. Noise in the neurons in her brain, in the amount of her muscle activations, and noise in all her sensing organs. Detecting a change of direction when under self-guidance is riddled with uncertainty. Even if an error is detected accurately, there is always some uncertainty in the estimates for how much to compensate. The math predicts there will be both over- and undercorrections: zigzagging to stay on course, amplified by just a bit of noise. This can be seen in Souman's plots of his participants wandering through the woods and across the desert. They zigzagged back and forth by as much as one hundred meters as they tried to keep on a straight line. That is a lot of slop! Because each new adjustment adds more slop, the errors accumulate, one bad adjustment after another.

All those errors don't cancel one another out. Assuming even a modest amount of noise, positional uncertainty becomes so large that a purely self-guided walker is almost guaranteed to become lost. Any sort of bias to the right or left and a circle is inevitable. You can test this yourself. Find a quiet large field, blindfold yourself, and try to walk in a straight line. Make sure you can't use environmental sounds to guide your directionality. You will soon be walking in a tight circle.

The only way out of this predicament is to incorporate knowledge of the world from which you can create a frame of reference based on something besides the self. Once the traveler can find a way to use the features of the world, the math for staying on course, even in the face of noise, starts to work out beautifully.[5] Zigzagging diminishes, errors don't add up, and those that occur can be corrected quickly. The risk of going in a circle disappears. It can take remarkably little evidence for a person to capitalize on a reference frame anchored in the external world. The subtleties of this skill were an inspiration for Charles Darwin, who was enthralled by the ability of native tribes to extract seemingly invisible cues from barren environments to guide their way.[6] He read about these skills in the detailed writings of the polar explorer Ferdinand von Wrangel and considered them evidence for a particular kind of instinct.

There are lots of directional clues that can be used if one knows where to look. One of the simplest is the sun. Although it slowly moves across the sky, at any given time of day it can still provide a powerful azimuth clue. Souman's circling hikers were readily able to stay on course through the woods when it was sunny out. Notably, there are enormous individual differences in how well people can use cues like the sun to keep track of their cardinal directions.[7] Just ask your friends. Cross-check their ability to point to true north against a handheld compass. You will be amazed by how many friends are guessing when not cheating by using a road or some other man-made reference. At the opposite extreme you will find friends who are spectacular at always knowing the direction northward. I am one of those people who always seems to know which way north is within 10 degrees or so. As the sun follows its slow arc across the sky, I am completely unaware that I am monitoring its relative position. I must be automatically following the shadows and light, updating an internal frame of reference organized like a compass with me at the center. While Darwin suspected this ability was an instinct, it may also be the case that the skill is something acquired. I might have unwittingly learned to do this in childhood or adolescence as I explored the woods near my home. Unfortunately, as I headed out on that cloudy day, without a sun to track, keeping a straight line was not going to be easy.

Despite my poor odds at finding my way by using the sun, I did have one small directional clue. Studying my map, I noticed that the plateau I was crossing sloped at a reasonably uniform grade over many miles. My route cut a diagonal path across the slope. I couldn't really see the tilt so much as feel it under my feet. By sensing how much the ground was tipping to the side I had another way to build a crude reference frame anchored to the world, rather than to me. It was sketchy, but, if I maintained the correct angle between the direction of the slope and my path, I might be able to stay on a reasonable course. Like the sun, this kind of reference frame uses relative angles. There is a particular kind of map that the mind can create for handling relative angles, arranged much like the lines at the top of a model globe, longitudinal lines radiating outward from a pole at the center. This polar map places me at the pole, with the slope and my direction plotted as two longitudes jutting outward. However, do the math and you find that once you add a little noise, a reference frame organized on a map of polar coordinates can be a bad idea for tracking one's location in an otherwise featureless world. Once off track, it is hard for a person to estimate where exactly he is and thus how much to correct. Better than nothing, but a risky strategy.

I trudged onward, albeit with an increasing realization that the farther I went across the opaque landscape, the harder it became to tell where I actually was. I was losing track of the desired angle of the slope. It felt like I was working against it a little too much, that I should descend more. I desperately needed to find some more clues. I peered about, searching for a familiar landmark. I had been through this area many years previously and hoped to recognize a feature that might establish a new kind of reference frame built from landmark recognition. Ideally, I could string a bunch of features together to get to my goal. Hikers in Wales and Scotland and New England have capitalized on this simple strategy by building long series of cairns, tall piles of rocks spaced a hundred yards apart, to find their way through dense fog. Unfortunately, all I had was the mist.

What I really wished for was a reference frame that could use the kind of details laid out on a map. When I started the day, I had studied my map at great length, slowly memorizing its features. In my head I created an entirely new stripped-down two-dimensional map laid out on a grid. It had my campsite, the portal on the cliff edge, and a line between them. Other features, like the location of easy-to-recognize mountains, were also plotted on my virtual map. My paper map was laid out under a grid of thin lines at right angles, little crosshatched markings organized like the streets of Midtown Manhattan. Similarly, there is neural architecture in the hippocampus of my brain that is actually set up in a similar way, laid out as a two-dimensional rectangular grid. Recordings of neurons in this part of a mouse's brain reflect where the animal is on the grid: they track the place of the animal on a square, gridlike map, monitoring its position.[8] The map is very much like the little one on an iPhone showing where you are at any given moment. The rectilinear grid does the best job of overcoming noise and allowing a person to relocate herself if she gets off track. This can be tested in the lab by placing participants in a virtual city. They can be tasked to build a 2-D square grid map in their mind by means of exploration and experience. Alternatively, they can be told to find their way by going from landmark to landmark.[9] When participants used a mental grid map as the primary reference frame, they were able to maintain a much better overall sense of relative direction no matter where they were on the map. They had a big-picture understanding. Those who build a reference frame out of landmarks alone were only good at pointing to the location of the last landmark.

A truly amazing property of the mind is its ability to combine all sorts of maps to create a hybrid reference frame for finding one's way. For example, the brain can augment the grid map with other evidence. It has place cells to track position on the grid, along with other neurons that keep track of the relative direction of the head with respect to this underlying map. This is like the little arrow on the iPhone map that shows the direction you are heading. There are also border cells that track the edge of the map; thus the map in your mind ends where your knowledge ends. All of this requires a sophisticated mechanism that can seamlessly combine information from one kind of evidence with another, resulting in a composite representation, rich in detail, of how a person is located and oriented in the world.[10] Unlike a bacterium constrained by a single simple self-centered reference frame, a frame that is defined entirely by chemical sensing that drags it up or down a gradient, humans can pick and choose among grids, landmarks, and other features, organized in a host of self- or world-centered frames of reference, and translate between them with ease. I would have been happy to have any one of them available to steer me true. Unfortunately, my place and head direction cells were running at idle. The landscape around me was so utterly barren that I had no feature to serve as a landmark or any information with which to construct a grid map. I was back to dead reckoning from an arrow drawn in the dirt, with a bit of directional fine-tuning coming from the feeling of the slope underfoot.

Although people are flexible in how they can use different reference frames and features to find their way, they also have preferences and inherent differences of ability. We all know grid people who give driving instructions by drawing a map and landmark people who will describe a series of turns.* The capacity to switch between reference frames is something that changes over time; young adults are much more flexible in switching between different navigation strategies.[11] The elderly will rely more on landmarks[12] and will stick to routes that they already know and avoid shortcuts, which would require a switch from navigation by means of sequences of landmarks to a 2-D rectilinear grid map.[13] Getting lost because of this reduced grid map way of thinking, particularly in middle age, is a potential harbinger of mild cognitive impairment, perhaps a precursor of Alzheimer's. For those who develop full-blown Alzheimer's, the problem of not being able to develop a good grid map becomes compounded by a deterioration in remembering beacons.

* As far as I can tell, the exception is in Italy, where there seem to be no turns. Everyone I asked for directions simply pointed down the road and said "*Vai sempre diritto*" (keep straight on).

The reference frame a healthy person prefers determines more than just his mode of navigation, it profoundly shapes the very essence of how he understands and interacts with the world. Mentally making a grid map with many fine details takes cognitive effort and it is a practiced skill. Skipping the effort of creating a mental map can lead to a myopic comprehension of the environment. The insidious tendency to use landmark navigation as a quick fix is exacerbated by modern technologies. Cell phones and GPS with their precise coordinates trick a person into thinking she has an understanding of the grid. But a pair of coordinates on a map does not make spatial knowledge. Spatial relations, particularly in natural settings, are absent on tiny mobile phone maps, denying the user a big-picture understanding of the way the terrain imposes risks, opportunities, or constraints on movement. One simply obeys the GPS or text instructions like a dog on a leash. I've come across many hikers who relied on text instructions as directional vectors, only to pay a big price. The instructions are simple enough: at the first junction turn right, at the second junction left, then another left, etc. This is beacon navigation at its most basic level. Imagine what happens when they park at the wrong trailhead or accidentally miss any one of the many junctions that serve as beacons. Deep in the back country I have been asked by confident hikers, "Where's the waterfall?" They are convinced there will be one just around the corner and are stunned when I tell them the closest one is actually a dozen miles away, in a completely different watershed. They don't even know they are lost and they have no alternate point of reference that is built bottom-up, from the landscape itself. Knowledge comes from the clever combination of all the reference frames, including the details of waypoints, directions, landmarks, and the entire grid.

It is easy to see how a variety of reference frames allows

a person to paint a mental canvas so he finds his way. Less obvious is the fact that reference frames are also essential for controlling the body so it can accomplish even the most basic goal-oriented movements. At the start of my day I had drawn a compass rose in the dirt. To do that, I started with a spot on the ground located on a frame of reference centered on a map of the world, which I then converted to a reference frame based on the direction of my gaze and where the spot fell on my retina. This location had to be adjusted based on the orientation of my eyes with respect to a reference frame centered on the orientation of my head, which in turn had to be translated into a frame of reference centered on my shoulder, where my reaching arm was anchored. The fingertip had its own reference frame that needed to be related back to the shoulder. With brain scans, each of these different reference frames can be readily distinguished in successive regions of a person's brain.[14] It is a dizzying cascade, with the brain transforming coordinates from one reference frame to another, all of which have to be revised constantly as the shape of the body changes. All just to draw a circle in the dirt.

While one neuroscience community studies how the brain enables spatial navigation across the landscape, a completely separate community considers how a problem like pointing the finger in the dirt is controlled. These two problems use very different kinds of maps: one built from features in the world mapped to the cells in the hippocampus and the other from features of the body mapped to multisensory neurons in the parietal lobe. And yet they both share the same fundamental theoretical problem: how to combine different sources of information into a coherent understanding of the body situated in the world so that a person can act. The shared theoretical problem is figuring out how the brain combines information from different maps, whether it is integrating evidence of head direction and location or relating the direction of gaze and the location of the fingertip.

The notion of a reference frame, and how one frame could be transformed into another, can be traced back to the 1870s, when there was a medical obsession with extracting bullets. At that time, most surgeons were convinced that it was essential to remove a bullet if their patient was to survive. There was fear of infection, of lead poisoning, and of migration. Beginning in Germany, surgeons developed remarkable tools like forceps and skilled techniques at getting bullets out of the chest, abdomen, and muscles. However, the brain posed a far more difficult challenge. One couldn't simply open up the entire skull and push things around. The surgeon had to get in and get out along the narrowest possible pathway. The advent of X-ray radiographs made it possible to get a general sense of where the bullet might be lodged inside the skull, but surgeons were still bedeviled by uncertainties of location due to image parallax.

The elegant solution came not from a surgeon, but from an artist and sculptor turned photographer, Gaston Contremoulins. He had grown up learning the crafts of his family, watchmaking and machining, and he had studied fine art. Unable to make a living in art, Contremoulins learned the new method of photography and got a job as a technical photographer of microscopy for the Paris Faculty of Medicine. In 1897, one of the surgeons at his hospital and a close collaborator, Charles Rémy, had a twenty-year-old patient who was blind from a gunshot wound sustained at age thirteen. The patient and surgeon were convinced that removing the bullet from the frontal lobes might allow the young man to regain his sight. Rémy made a hole in the patient's skull and tried to scoop the bullet out by running his finger in the space between the brain and the skull just above the eyes but gave up after an hour of poking around. Contremoulins recognized that a couple of X-rays alone would not provide enough guidance. They needed precise geometric triangulation if they were going to calculate the exact location of the bullet.

Using his skills as both sculptor and machinist, Contremoulins created a rigid box that could be fitted onto the patient's head and held rigidly in place with screws and metal brackets. The sides of the box created a solid frame of reference to which an X-ray film could be mounted on one side and two X-ray sources (Crookes tubes) could be mounted on the other. Each tube made a distinct shadow on the X-ray. This frame, which he called the *"compas repère,"* or reference compass, made it possible to overcome any uncertainty about parallax. After detaching the *compas repère* from the patient, he attached two wires inside the box, spanning the Crookes tubes and the shadows on the X-ray. The point of intersection was the exact location of the bullet with respect to the box. The problem of course was that he had located the bullet relative to a reference frame defined by the box, the *compas repère.* The surgeon couldn't do anything if the patient still had a box on his head. Contremoulins's deep insight was finding a way to translate his X-ray-centered reference frame to a skull-centered reference frame. His solution was ingenious in its simplicity. He created a second reference frame, which he called *"compas schéma,"* or planning compass. It was a set of metal arcs that could be attached to the patient's skull. The critical step that made the entire process possible was to attach the *compas schéma* at exactly the same spots on the skull as the *compas repère* had been attached. With this, the skull served as an immutable third reference frame to which all the others related. Having made the attachment, he could slide the arcs of the *compas schéma* across one another until they matched the angles of the wires; now the location of the bullet and the desired path to get to it could be planned with respect to the skull.

After practicing with an old skull, Rémy and Contremoulins took the blind patient to the operating suite and extracted the bullet with complete success (although the patient's vision was not restored).[15] Contremoulins had invented stereotaxic surgery, a now-ubiquitous neurosurgical technique. While the modern neurosurgeon will use fancier X-ray imaging and mathematics to calculate angles rather than interlocking arcs, in the end, she still relies on this fundamental idea. Like Contremoulins's, her solution relies on the skull acting as an anchor, a master reference frame to which all others must relate.

Stereotaxy works because different frames of reference can be literally bolted to the head. The skull is the master reference frame with which all others must agree. But the reference frames that a brain uses to understand the world are not bolted. Nevertheless, there has been a strong intuition since Contremoulins's time that the brain might have strong preferences for using a single reference frame, like the skull, as a fixed map to which all others are related. This idea became particularly prominent for arm actions such as pointing and grasping. No matter what sense was being used to help guide the limb, whether it was sound, vision, or the sixth sense, all that information would first get translated into a master reference frame. By the 1990s the hunt was on in neurophysiology labs to find what that frame might be, and for a while it seemed like the eyes would win out. Whatever direction the eyes were pointing and perceiving would be the anchor, like the skull in stereotaxic surgery. Any other senses would have to first be related to what the eyes saw. This seemed plausible; it feels right to be anchored to what is seen in the external world, a comparatively constant physical entity with clear spatial properties. Vision has the added value of great spatial precision relative to the other senses. Bolstering this idea, recordings of neurons in the brains of animals as they looked at or reached toward targets showed that there is an enormous amount of real estate dedicated to tracking targets exactly in terms of where the target image lands on the back of the eye.[16]

At the same time, it is easy to find situations where the gaze can't be the master reference frame, and if it were, it might even get in the way of performance. There is a wonderful video of Michael Jordan in his prime, with a huge smile, making a free throw with his eyes shut. By closing them, he could anchor himself to the world using a reference frame other than his eyes. Look closely and you will see that he held his head quite still and used head orientation relative to the basket as his primary frame of reference. There are also videos of people shooting one hundred baskets or more in a row with their eyes shut. They never open the eyes between shots (someone hands them the ball before each throw). They use the same solution as Michael Jordan: they keep the head perfectly still with respect to the basket. The trick of using the head rather than the eyes as a stable reference frame has now been adopted by many basketball players to improve their free-throw percentages. Basketball teaches us that there is no absolute master reference frame for physical actions. Instead, there is a constant switching among all kinds of reference frames, fluid and hidden and driven by many surprising trade-offs.

This ability to switch between reference frames is something we studied with brain imaging in my lab. Pierre-Michel Bernier, a postdoctoral fellow at the time, was able to identify one of the sweet spots in the parietal cortex where this kind of elegant pivoting from one reference frame to another takes place. He showed that in addition to areas that seem to be dedicated to one reference frame or another, there are also intermediate areas that can do the flip-flop, switching back and forth across frames depending on the demands of the task.[17]

—

As I found my way, I realized I was drawing on another obscure force to construct yet another azimuth to help me track my course: gravity. It is a perfect directional cue for an earthbound traveler: always present, immutable, and easy to detect thanks to the otoliths located deep inside the inner ear.[18] Otoliths are cells with little hairs bulging up into a bath of fluid. At the tip of each hair is a block of rock-hard calcium that will bend the hair in relationship to gravity and other head forces. The signals from the otoliths of each ear track gravitational force. However, to create an accurate prediction of the direction toward the center of the Earth, signals from both ears need to be combined and calibrated to each other. This requirement is readily apparent in a person suffering from a viral infection that disrupts the otoliths in just one ear. The miserable patient will lurch sideways, stumbling against walls as if gravity itself were drawing her sideways. Losing a sense of the down direction is also a major problem for two-thirds of all astronauts, who develop nausea, lethargy, and motion sickness while in space. Fortunately, for most of us the otoliths stay nicely tuned to gravity. You can easily tell if everyone around you is tracking the down direction, whether they are urban strollers, athletes in competition, or commuters heading for work. Is anyone walking with a tilted head? On the contrary, everyone maintains his or her head almost perfectly level to optimize performance of the otoliths. They are all unconsciously tracking gravity, establishing an azimuth that provides a critical pivot point for navigation.

You can observe how necessary it is to keep the head steady by watching a toddler totter through his home. The route he takes, while not random, is based mostly on self-propulsion and dead reckoning, a sort of point-and-shoot strategy with the head bobbling about. The abilities to stabilize the head to gravity and, just as important, to anticipate which direction to point the nose as walking direction changes aren't fully developed until around eleven to thirteen years of age. Only in adolescence do locomotor trajectories become smooth and consistent.[19] Brain scans of patients who have suffered cerebral stroke reveal that the internal sense of verticality provided by gravity requires an intact posterior insula (a deep area of parietal cortex known to integrate gravity, the visceral senses, head orientation, body motion, and a grid map of allocentric space into a composite representation).[20]

Gravity doesn't just help in navigation and walking. It also aids eye-hand coordination. When I pointed in the dirt, the frame of reference used for planning movement of my arm was being tuned in part to the gravitational vector. This is easily detected in laboratory experiments by having a participant tilt her head sideways while pointing at a target. Her accuracy will be significantly impaired.[21] "Graviception" is critical for many sports, including soccer.[22] Kicking a soccer ball requires stable support of one leg so the other can make precise contact with the ball. Soccer players are thought to develop better internal representation of verticality than nonplayers.[23] Competitive soccer players in particular are better able to withstand experimental disruptions of cutaneous touch, muscle position, and joint angle without losing their sense of balance compared to less trained players, suggesting they are better able to incorporate gravitational information to maintain a sense of verticality.[24]

The recognition that there is no canonical or master reference frame has led neuroscience to adopt a flip-flop model, in which any reference frame can be the ad hoc master, given a particular situation. I could use either a grid map or landmarks as a reference frame. Similarly, I could plan an arm movement with respect to the fingertip or, if I was holding a full cup of coffee, with respect to the orientation of the cup to keep it from spilling. Beginning in the 1990s, a number of computational neuroscientists began to realize that even the flip-flop model would fail under many conditions.[25] Things get particularly messy whenever there is motion involved. To get an intuition of this problem, imagine a bicyclist quickly cutting across your direction of travel. The front wheel of the bicycle has a shiny reflector close to the rim. If you hold your eyes still, the image of the reflector creates a path on your retina that is moving up and down and also sideways, projecting the image of a jagged wave. If you used a frame of reference centered on the eye, you'd perceive this wavy motion. However, you don't see that at all. You see a reflector going around and around in a circle on a wheel. To make sense out of this, the brain has to find a way to go somewhere in between the wave imprint on the retina and a reference frame centered on the wheel. Not the bicycle. Not the rider. Not the path. The axle of the wheel. Projecting the information onto the retina or the wheel alone won't work. But in an abstract in-between reference frame, the transformation of the reflector moving as a perfect circle is no problem at all.[26]

One way this in-between frame can be described is with something mathematicians call "vector fields," a very abstract way of describing relative motion through space.[27] While this is hard to imagine, it is actually the sort of computation that is straightforward for neurons to perform. Hybrid reference frames provide a graceful way to combine all sorts of evidence into useful models yielding far more detailed knowledge of both the organism's state and its environment, either of which could be in motion. Hybrid reference frames like this do more than just make sense out of visual motion. They are integral to just about anything we do when our body is in motion.[28] A good test of this is grabbing for the brass ring on a merry-go-round. Up and down, round and round, the rider has to adjust the timing of his reach and grasp at just the right moment.

The value of hybrid reference frames becomes clearest in peak athletic performance. Speed skating is the fastest human race on an unbanked track, with Olympic athletes routinely reaching speeds of thirty-five miles an hour. The fact that they are slipping around on ice makes the role of balance and an ability to integrate a gravitational vector into the reference frame obvious. In addition, there is a lot more to skating than keeping upright and tracing an oval. These athletes go fast. Sharp turns impose a completely different set of forces than do straightaways. The skaters struggle to maximize power and simultaneously tune their centripetal acceleration so as to avoid blowing out of a curve. Without even knowing how they do it, they create a hybrid reference frame that is centered out in front of them. They create a virtual path to follow that is in theory the optimal solution for integrating the senses.

Joan Vickers at the University of Calgary wondered if differences in how athletes defined this virtual line influenced their performance.[29] She installed a combination eye tracker and video camera on each athlete's head. Not surprisingly, all the skaters, whether they were elite or nonelite speed skaters, spent their time gazing ahead rather than at the sidelines or down toward their skates. Looking more closely, she could determine where their gaze fell on the ice. The elite athletes spent far more time picking the line—that is, looking at the tangent point on the arc they would be skating or, better still, looking just inside this arc. This inner spot is the ideal frame of reference for balancing all the competing forces of motion. The nonelite athletes spent more time simply looking straight ahead (rather than down the arc) or looking outside the arc. They had yet to create an ideal hybrid reference frame essential to attaining the next level of performance.

Another advantage of hybrid reference frames is that they provide a very effective way to deal with noise and uncertainty. There are elegant computational models that explain how noisy neurons can work in groups to combine different kinds of information into a common reference frame and, just as important, to adjust the influence of the different neurons so that they will rely more or less on different senses depending on the relative noise.[30] As I walked across the plateau, I was estimating the slope of the terrain, trying to cut across it at just the right angle. Normally, I might rely mostly on vision to judge the slope. Immersed in the foggy haze, I was forced to rely mostly on what my feet felt. This adjustment was automatic, easy, and unconscious. From experiments in the laboratory, we now know that for a lot of actions, like reaching and pointing, the brain can do a nearly perfect job at trading off the different senses to build up the most accurate hybrid reference frame. Just ask a person to reach toward a target but play tricks on her so she is less sure of what she is seeing or feeling. Her brain will smoothly shift reliance toward whichever sense is providing the best evidence.[31]

Most of the time, hybrid reference frames operate under the hood, efficiently integrating disparate senses into a coherent structure. Much of this is innate, designed through evolution to help us solve basic tasks or to make sense of the world. However, this structure can also be hijacked, creating wild misperceptions of the environment if the expected relationships among the different senses are violated. The Hong Kong Peak Tram is a train that takes tourists to the top of Victoria Peak for breathtaking views of the city. It is a steep climb (average grade 18.7 degrees), roughly three times steeper than the steepest highway. Riders sit in a regular sort of train car, so that they are tipped backward as the train crawls up the hill. As they look out at the many high-rise buildings of the city, the passengers are often struck by an illusion that the buildings are tipping over.[32] It is a powerful visual illusion, with the buildings leaning about 30 degrees and away from the observer—a crazy amount of tilt. They hold their hands out at a 30-degree angle to get a sense of how far over they think the high-rises are tipped. Clearly, something odd is going on in the observers. The interesting thing about this illusion is that no single sense is being tricked. The investigators who described the illusion tried to eliminate it by systematically removing potential sources of sensory conflict. They put a cardboard box with a hole in the side on an observer's head, so the observer couldn't use the windows or train car as a reference frame. All he could see was the distant city. The tip was only a little less severe. They had the person sit in a forward-tipped chair, like a funicular, so that his body was upright with respect to gravity. Only a little less tip to the building illusion. They had him stand upright. Only a little better. As they started to combine these manipulations (stand with a box on the head) they finally started to make some progress at removing the illusion of tipping buildings. In other words, it was the amalgamation of different senses into a distorted hybrid reference frame that led to the illusion. The illusion occurs because this kind of tipped, passive motion wasn't around while our species was evolving, and our hybrid reference frame expects there to be fixed relationships among optic flow, body motion, and gravity. Violations of those expected relationships change what a person thinks is vertical.

On my walk, time was the ultimate reference frame. Based on the number of hours I had been traveling, I deduced I should be getting close to the cliff edge and, I hoped, the lone spot where I could get down. Otherwise, I was walking in circles. I peered through the clouds, and in the far distance I could just make out a scraggly rock, shaped like a giant statue of the hunchback of Notre Dame. Smiling with relief, I remembered this weird formation from a trip I had taken through this area previously. It was close to the drop-off point. Long since forgotten, this landmark was instantly recalled. It was only about 30 degrees off my current heading. I updated my mental map in a blink and made a beeline for it.

7

Learning to Solve Problems

I don't think she's smarter than most bears.
I think she's had more time to learn.

—BEN TABOR, New York State wildlife technician commenting on Yellow-Yellow, an Adirondack black bear[1]

AFTER MANY DAYS living up at altitude, sleeping in barren alcoves tucked into the cliffs, mushing through cutting winds, and weaving among rocks, rocks, and more rocks, I looked forward to a night in the comfort of a thick pine forest as a fabulous luxury. The wind quietly rustles high above in a canopy of evergreen needles, the tent is laid out upon a soft forest floor, and a sweet smell of sap permeates the air. I had dropped into a delightful gorge with a plan of climbing out the other side the next morning. Unfortunately, my route would transect the single largest hiking trail in the range, a massive artery snaking the full length of the gorge. It was heavily used—the dirt was ground to dust like cake flour. Shuffling pack trains of mules kicked it up, leaving a dense cloud that clung to sweaty hikers toiling along the same thoroughfare. I was relieved my path was perpendicular to theirs.

By the time I got to the trail crossing it was dusk, and I needed to settle in. As the main thoroughfare, it was also the epicenter of bear activity. Years ago, the Park Service had planted bear boxes at campsites in the gorge to help packers keep possession of their supplies. Even though I was hauling a supposedly bear-proof food container, I decided I would be better off camping near the dusty trail so I could avail myself of one of those heavy-duty metal bear boxes. I found a nice-enough spot reasonably far from the dust bowl, overlooking a swift stream cascading down a terrace of small ledges. Although small, the stream represented the headwaters of what would eventually grow into one of California's biggest rivers. I made camp: not much more than setting up a tent, pulling out my bedding, and cooking a simple dinner. My last chore was protecting my food. Shoving my plastic food container into the metal bear box, I could make out neat little scratches by the handle. They weren't the desperate scratches of a dog pawing for some food. They were all neatly spaced, about as wide as the fingers of an enormous person with fantastically long fingernails. The scratches left me feeling queasy. Human hands weren't big enough nor were human fingernails sharp enough to make those etchings. A bear must have clawed the door.

The scratches reflected a long-standing technology battle between humans and bears. Designing any sort of animal-proof food container has been a chronic problem for people for as long as there has been cultivation. A food container has to be tricky enough to stump a hungry animal but not too hard for a person to figure out. When considering the various bear boxes in our national parks, one notices an escalation of complexity, each new design informed by a presumption about the limits of bear cognition. Many of these bear boxes are embarrassingly naïve in construction. The bears have defeated each new model with their ability to learn new ways of problem-solving. It turns out that problem-solving and learning—among people and bears—are just different expressions of the same underlying cognitive machinery.

To produce those scratches in the metal, a bear had to overcome any trepidation it may have felt about messing with the box in the first place. It is a foreign object. It could be a trap or something that causes pain. Before problem-solving can be done, an animal or person must be willing to engage. There are enormous differences in diffidence between species. Bonobos are wary of novelty,[2] whereas orangutans are like small children, spectacular explorers and confident manipulators. There was a happy time fifty years ago in the national parks when bears were extremely cautious of humans and their trappings. No longer. I had learned this the hard way when I was a young climber. One night I managed to bed down in what turned out to be a thoroughfare for Yosemite Valley bears. I was in the middle of a trail connecting their sleeping den and a crowded food-laden campground. The paradox of a 450-pound black bear is its stealth. It can approach a sleeping person with scarcely a sound. What gives it away is its breath. I was lulled from sleep by an overwhelming stench, like that of a hot alleyway dumpster. When I opened my eyes all I saw was a massive brown muzzle sniffing my face. I froze, hoping that if I pretended I was invisible or looked like I was dead the bear might just go away. There wasn't a whole lot else I could do anyway. Only my head was sticking out of my sleeping bag. The rest of me was pretty much encased in a down straitjacket. Fortunately for me, Yosemite bears are professionals when handling humans. The bear used its muzzle to shove my head out of the way, then slipped his tongue into a pocket of my down coat, which had been serving as my pillow, and consumed a little piece of chocolate I had left there. It was no bigger than a postage stamp, but he found it. Satisfied, he silently lumbered off, a giant black shadow jiggling through the trees.

By the early 1970s, bears had learned that green-and-white metal Coleman coolers were loaded with goodies. They were beginning to problem-solve in the simplest way. They had learned that a particular stimulus was associated with a great reward. The rangers countered by advising car campers to throw a tarp over their coolers so the bears wouldn't recognize the stimulus. The bears figured out the cooler disguise almost instantly and learned to rely on their noses. The rangers offered new advice: put the coolers in the car trunk. Their assumptions had progressed: bears should be wary of this large human object, and they would be unable to approach. Soon, this advice had to be customized for owners of Volkswagen Bugs. The bears learned that if they climbed on top of a Bug and jumped up and down, the bouncing would pop open the trunk and reveal the cooler. By the late 1970s, they were willing to approach any car and progressively discovered a variety of techniques for smashing through rear windows. They would crawl inside, and if it was a sedan, they would rip apart the backseat until the trunk was accessible. I owned one of those cars.

Bears were doing more than confronting the problem of recognizing a cooler at a campsite. They were messing around, experimenting, trying out different actions requiring an internal drive to keep working at the problem. Chewing through the rear seat of a car takes a lot of effort, but for most carnivores, it takes only a bit of hunger to motivate them to keep working at a problem like this. The smells coming from the trunk are more than a sufficient motivator.

To measure how hard different species will work at this kind of problem, Sarah Benson-Amram recorded the ability of thirty-nine different hungry carnivores, including big cats, canines, and bears, to open a puzzle box loaded with their preferred meat. She used a metal box whose sides were composed of steel bars, a miniature jail cell as it were, with a hinged door held shut by a sturdy latch on one side. The animals could see and smell whatever was in the box, which was carefully sized to match the proportions of their paws. All the beasts had to do to get the food was slide the latch handle over and swing the door open. Their efforts were video-recorded and their performance graded. The main takeaway from the study was that the size of the brain relative to the size of the body correlated with the chance of success. But that was only one way to look at the data. There were outlier species, like bears and big cats, that did strikingly well. All of the carnivores had clearly achieved a stage of cognitive development that the psychologist Jerome Bruner had proposed as an early milestone for problem-solving by children:[3] the intention to work, the ability to sustain effort at solving a problem. The bears worked at the latch problem for two to ten minutes. In humans, this willingness to do work appears at an extremely young age. Watch a ten-week-old infant in a crib; hang a visually stimulating mobile over the infant and it will double or triple its spontaneous kicking for a prolonged period to get the mobile to move.[4]

Bears will work to get into a food container for far longer than ten minutes. A few years ago, I tried out the latest ultralight bear-proof food container for backcountry travel. On first glance, it was a throwback to the food sacks used in the 1950s, a bag that could be tied high up a tree. Probably one of the oldest safekeeping solutions ever devised by humans was to place food out of reach. What made this sack special was the fabric, a military-grade Kevlar tied shut with an unbreakable half-inch braided pull string. The bag is a lot lighter than plastic or metal containers and increasingly popular with fast hikers obsessed with keeping their pack weight down to an ultralight minimum. The idea behind this tough sack was that a bear, even if it could reach it, couldn't pull it out of the tree or rip it open. The design wrongly assumed that a bear would lose interest if it couldn't pull the sack down from the tree or rip it open. The very first night I tried using the sack, two bears, one big, one small, showed up. The big one was able to climb high enough to reach my brand-new food bag, stuffed with a week's worth of my carefully chosen favorite foods. The designers were partially correct. The bear gave up clawing at the bag, which held firm. What the designers hadn't fully counted on was the bear's willingness simply to chew on the bag as if it were a giant marshmallow. After more than three hours of the bear's salivating and masticating, the contents of the sack had become a sludgy goo that the bear sucked with determination—a noisy, sloppy affair, interspersed with occasional roars as the larger bear chased the whimpering adolescent away.

Bears aren't unique in their willingness to work at a problem; voles, squirrels, and marmots will also chew for hours to get through a tent or pack to steal food. We fail to recognize the importance of this willingness in people, who also need to be able to stick at a hard problem until it is solved, whether they are a surgeon grappling with a difficult case in the operating suite or a plumber under the house with a bad pipe. Given the bears' resolve and willingness to work for food, the Park Service had to go back to the drawing board and invent harder problems for them to solve. But solutions were not obvious. The first wave of metal bear boxes, huge steel containers cemented into the ground, assumed that bears lacked another cognitive ability that develops in children, facility in what Bruner called the "assembly of constituent acts": the ability to put together a series of actions to achieve a goal. The baby kicking at the mobile, the bear chewing on a food bag or ripping apart a car: these are not sequences of different acts. A two-step action like pulling a latch to open a door is a better example. The first generation of bear boxes had this simple kind of design. In online videos of Sarah Benson-Amram's zoo research, one gets to see how each carnivore manages to open the puzzle box in the final few seconds before the latch is flipped and pulled so the door swings open. Seven of ten bears (grizzlies, polar, or black bears) were able to solve the latch problem, to assemble constituent acts, just as they quickly learned to do throughout the parks. What is missing from Benson-Amram's videos is a sense of how much guesswork or dumb luck was involved before they flipped the latch. When the carnivores looked at the latch, did they understand something about the physical causality inherent in the design of the latch? Or were they just fumbling randomly with it until something useful happened? This simple distinction underscores one of the most important conceptual leaps in problem-solving: trial and error with a bit of dumb luck, or a mental model informed by insight?

The next generation of bear boxes had a steel door closed by a heavy latch handle that was covered with a metal hood. This required the opener to reach inside with the fingertips, palm upward, push up on the latch, and then swing open the door. To do that, a bear would have to assemble constituent acts (push the latch then lift the door) with an underhand reach, which they rarely practice in the wild. The designer of that box had added a physical problem for the bear to solve, having assumed that bears lacked not only dexterity but the ability or patience to learn to form an underhand grip to force the hidden latch. Bears have proved these assumptions wrong. They have long since cracked this kind of a box. In doing so, they blurred the distinction between what we think of as problem-solving and what we normally call skill learning.

To appreciate how bears learned to solve this physical problem, it is helpful to ask how you might program a computer to teach a robot to open the foregoing kind of bear box. A deep insight into how to do this was offered by Andy Barto at the University of Massachusetts Amherst. Barto is a computer scientist and a world expert in writing algorithms that are used to train robots to perform very specific and precise actions, like those cutting, welding, and riveting robots on an automobile assembly line. I was fortunate to work with him on a project that studied how the brain detected and corrected for errors in fast precise movements. It turns out that the algorithms he developed in computer science are very similar to key ways that the human or bear brain learns, driving us toward motor perfection. Barto and his colleagues' solution is generally referred to as "reinforcement learning."[5] Imagine a coach who is trying to teach you a new skill. You are the actor. The coach is the critic. After each attempt, you commit a certain amount of error. Your coach doesn't comment on why you made the error but only on whether your attempt is different from what he expects from you, for better or worse. If your next attempt is a lot better, the coach offers a loud positive comment; if it is pathetic, a loud negative comment. If you are doing what the coach regards as typical for you, he is silent. Such a coach might drive a person crazy because he never tells you how to get better. He simply acts as an amplifier that reflects how much performance has changed for better or worse. You, the actor, have to decide how you will change your behavior in response to deviations from what the coach expects. You are in the hot seat, compelled either to exploit certain behaviors or to explore new ones. In actuality, you, the actor, and the coach, the critic, are inside your brain, each monitoring the behavior of the other. And the whole thing works only if you are creative, if you can try different behaviors as you learn. This is what is going on when a bear jumps on a VW Bug to open the trunk or shapes its hand to open a hooded latch, when a computer algorithm leads a robot to weld perfectly in a factory, or when a baby holds a spoonful of food without dropping it.

The beauty of reinforcement learning is that it allows for exploration; the actor can test all kinds of actions to solve a problem. It works best when there are a clear-cut goal (food in a box), a reward, and plenty of opportunities to try different solutions. The actor is also learning prediction, discovering what will happen next as a consequence of its action. As the motor system of the bear learns to adjust hand posture, it is predicting how the digits will reach the latch. It forecasts what will happen in the very near future. Forecasting is essential if the bear's action is to be successful. The body must anticipate what comes next, not just what it is doing in the moment.

Forecasting and prediction in the motor system become wonderfully clear in the hands of concert pianists as they perform difficult arpeggios up and down a keyboard. To make this flurry of notes, the pianist's hand has to be constantly repositioned. John Soechting and his students measured these adjustments formally by recording finger and wrist positions as pianists did their thing.[6] They found that the wrist moved in a way that was anticipating locations over the keyboard that would be optimal for the upcoming finger movement. And the forecast wasn't distant, as imagining the hand position too far into the future would undermine performance. Only a couple of notes on the keyboard were anticipated, just enough to allow the arpeggio to be sustained smoothly over a long passage. Similarly, bears smoothly adjusted posture and limb orientation in anticipation of what the bear-box latch required. It is easy to take this physical kind of problem-solving for granted. It is so automatic, smooth, and precise we don't think about it. But motor forecasting is inherent in almost every physical problem that we solve. Whether a person jumps over a pothole or lifts a heavy object, his body makes a flurry of preparatory muscle contractions that serve to maintain posture and form, reflecting the dynamism of physical problem-solving.

Reinforcement learning is particularly good for training robots to do one thing, like shoot a rivet at an exact location or weld two panels along a perfectly straight line. The robot is extremely sensitive to the details in which learning takes place. Change the nature of the problem just a little bit and the robot has to start from scratch and relearn what to do. Fortunately, the human brain can transcend this limitation. It has a built-in mechanism so that it doesn't necessarily care too much about the exact situation in which learning is taking place. It can generalize in ways that artificial intelligence is nowhere close to approximating. The bear can generalize as well. The skill it learned in opening hooded latches on bear boxes can be used for all sorts of handles. On returning to my car after trips in the mountains I no longer find broken windows and chewed backseats. The windows of modern cars are too thick to break. Instead, I find scratch marks in the dust around my car door handles. Looking around the dirt parking lot, I can see that a bear has tested every car door in the lot. Although each handle requires very different physical movements, one can readily see how the techniques acquired in opening the bear box were generalized, allowing a bear to access any unlocked car. Here again, by generalizing a newly created skill, the bear has a compendium of actions that serve as potential solutions to an even larger array of physical problems.

Neuroscience has described with great detail the brain circuits that allow reinforcement learning to happen, areas that operate as the actor and critic in learning. The circuits for the critic project from the cortex, through deep nuclei (the basal ganglia), and then loop back to the original cortical areas, forming a unique kind of feedback that can shape cortical processing. The loops are ubiquitous, allowing reinforcement learning for every type of cognition, not just motor skills. We know that the main neurotransmitter that signals discrepancy in expected reward is dopamine. Understanding generalization and how it is achieved by the brain is one of the most important theoretical challenges in neuroscience today. There is emerging evidence that the deep nuclei play an essential role in tracking the context in which learning takes place, a necessary process for generalization. Nevertheless, the rules for determining when there will be some generalization but not too much are still vague.

The bear's mastery of handles and levers forced the bear-box designers to up the ante with new models. For backpackers, one of the solutions was a large plastic cylinder with a lid that operates much like the familiar childproof cap on a medicine bottle. The lid has to be twisted as a tiny tab is simultaneously pushed. The container is large enough for a week's worth of food. To open it, one must hold it in place with one's feet while twisting the lid and pushing the tab with precise timing. If the feet let up, the whole thing will rotate and nothing will happen. Opening the container requires more than a series of constituent actions; the actions must be executed in a prescribed order. The container's designers assumed that this level of sequential coordination, combined with a high degree of manual dexterity, would be beyond a bear's capacity. Wrong again. The designers were shocked when, in the mid-2000s, a black bear in the Adirondacks named Yellow-Yellow learned how to open the container. She would bite the lid with her massive jaw, position her incisors against the childproofing tab, and simultaneously brace the jar section on the ground as she rotated the top off. Here is a clear-cut example of how bears can readily learn a short sequence of actions in a specific order.

Humans begin to learn exact, dependent sequences to solve problems at a very young age. This is clear as you watch an infant girl as she learns to feed herself. Put a spoonful of applesauce in front of her and she will reach out and grasp it, usually with the palm down and fingers wrapped nicely around the spoon handle, reflecting a growing capacity to anticipate arm posture and finger shape. More important, watch what happens if the handle of the spoon is oriented so that when she grabs it, the sauce end of the spoon is out by the pinky, not by the thumb and index finger. In this configuration, getting the applesauce to the mouth without spilling it is almost impossible. Try it! Rachel Keen spent much of her research career watching kids spill food, learning how they reshaped their mind as they discovered solutions to this very physical kind of ordered planning. To prevent a spill, the infant needs to discover and perfect other strategies. One of the first strategies that she can learn is to pass the spoon to the other hand. This solution teaches her a simple two-step prospective plan. The downside of this plan is that passing the spoon can be clumsy and the applesauce can spill. Eventually she will learn another kind of two-step prospective plan. She will first inhibit the preferred hand and then use the other hand. She will start learning to do this two-step planning as soon as she can hold something in her hand, at roughly nine months of age.[7] Keen found that by two years of age, infants are able to use one of the correct types of grips at least 75 percent of the time.[8] Not bad at all. However, this level of success was accomplished only after thousands of hand-to-mouth repetitions.

Sequential problem-solving is vastly more difficult to learn than simple actions that are acquired via reinforcement learning—actions like jumping on a car or pushing on a latch. Undaunted, Andy Barto and a colleague, Matt Botvinick, developed a host of sophisticated algorithms to train robots in essence to learn what the two-year-old was learning as she tried different methods to get the applesauce into her mouth. The challenge was to find a computer algorithm that could train a robot to achieve any task involving a sequence of actions. In the early days when they were working on this problem, I visited Barto's lab and was met by a seven-foot-tall robot, like

Robby the Robot from old 1950s science-fiction movies. As I walked into the room, it suddenly spun to face me, causing me to jump back. Its anthropomorphic qualities made me shudder, and that feeling was no less intense after I noticed that the robot was bolted to the floor. Andy's students were trying to get the robot to learn a simple enough task. In front of the robot was a table with a peanut-butter jar sitting on its side, within reach. They wanted the robot to learn how to pick it up and place it upright, with the obvious, bright red lid toward the ceiling. That would be step one of the more difficult problem of opening the jar and making a sandwich. The robot had rudimentary hands and arms with humanlike joints. Like the baby with the spoon, the robot grasped inefficiently the first time; its subsequent attempts were successful only if the robot used an underhanded grasp or if it passed the jar to the other hand. It had to recognize when to follow multiple steps and what those steps ought to be. To make the learning process akin to that of a two-year-old, the programmers weren't allowed to teach the robot any specific movements or even to show it what the goal was. Like the infant, it was programmed with hand preference and given a driving force motivating it to explore its environment. And if the jar ended up in the correct position, the robot got the digital equivalent of a gold star. That's it. The hard problem, the one that Barto worked on for many years, was figuring out how a string of actions could be combined in an orderly way. The baby with the spoon and the robot with the jar don't get input about success until they complete multiple actions. If the child is successful, is it because she grasped the spoon with one hand or because she passed it to the other hand, or both? Which bits of action are the basis for success or failure? Barto and his colleagues figured out ways to nest multiple learning processes within one another, the result being hierarchical reinforcement learning. The human brain does this too, relying on learning loops within other learning loops. Remarkably, this kind of learning is almost trivial when performed by the human brain, but analogous architectures are still extremely difficult to emulate in computer science and remain a major barrier that must be surpassed if robots are going to learn to perform behaviors similar to what a baby or bear can do.

Reinforcement learning and its more sophisticated hierarchical cousin are increasingly referred to as "model-free forms of learning." Simply put, the animal or person doesn't need to do any thinking to get to the solution; there is no mental model. Through trial, error, and feedback over thousands of attempts, a solution is eventually found. The advantage of this kind of learning is that once it is achieved, the brain has built into its circuits a solution that is almost hardwired and that can be reused automatically, with minimal cognitive effort or cost. Activities like putting your clothes on in the morning can be done with no thought. The downside of this kind of learning is the thousands of trials it takes to get things right. Parents may recall how long it takes for children to learn to dress themselves. The alternative to model-free learning is, of course, what is known as "model-based learning," where some kind of thinking is involved. While putting clothes on can be quick and automatic in the morning, deliberation over what to wear to a party is an entirely different matter.

Whether or not animals can solve problems using model-based methods is debatable, since we don't really know what it means for animals to "think." One basic mode of thinking is the ability to create a mental image for something not present. Most of the problem-solving that animals have been tasked with involves food rewards that can be seen or smelled, safeguarded in puzzle boxes or car trunks. To get at the food requires manipulation of the box or car, but there is no need for a mental model separable from the food. What if the solution to the problem is indirect, unrelated to the container of the reward? In the 1920s, Wolfgang Köhler devised a simple experiment to test if animals could build a more abstract cognitive model in which the solution to a problem was not obvious.[9] He challenged a hungry chimpanzee by hanging a banana from a rope, well out of reach. There wasn't a ladder or tree available and there was no access to any of the rope work. The chimpanzee had the necessary cognitive capacity: it gathered a bunch of wooden crates that were spread around its compound and stacked them into a makeshift ladder, climbed on top, and nabbed the banana. A simple mental representation of a ladder/tree-shaped object under the banana provided a way for the chimp to find a solution without first trying thousands of actions using reinforcement learning. However, the gain of efficiency in this model-based learning isn't free—it takes cognitive resources (*i.e.*, smarts) and mental effort.

As for bears, do they have what it takes? Years ago, I was "lucky" enough to observe a bear engaged in model-based problem-solving in an experiment that Köhler would have loved. I was camping in the Little Yosemite Valley, an area tucked behind the monolithic Half Dome. Back then, there were no bear boxes or fancy containers. To protect my food, I did exactly what was in the Boy Scout manual. I climbed up two trees and suspended a taught line between them. Then I threw a second line over the high line and hung the food in midair, about fifteen feet off the ground with the tail of the rope to the food bag tied high in the trees. There was a gorgeous full moon and I slept in the open, taking in the light illuminating the back side of Half Dome. At around two a.m. a hefty black bear woke me up as he knocked an empty cooking pot around. Then he caught scent of the food up above. I made some noise; he gave me a cold stare. Realizing he wasn't about to go away, I sidled up against a downed tree to watch what he would do. Ignoring me, he unsuccessfully searched for any lines leading up to the food: the direct model-free solution. Then he did an amazing thing. He began dragging downed wood and brush into the campsite, where he piled them up right below the food. The pile was messy, but there was enough stability for him to climb on top and reach toward the food bag. I was stunned. The bear was behaving like the chimpanzee in Köhler's experiment. I was relieved when the bear failed to make a pile tall enough—there weren't enough logs around. Sadly, my relief was short-lived. Here was a highly motivated problem solver. He stepped back and assessed the situation. In the bright moonlight, he conducted a direct assault. He eventually spotted the rope tied to the food bag. He tracked its course to where it was tied high up a tree. Resorting to something he had probably learned through reinforcement, he climbed up the tree and with a lunging reach pawed the rope, yanked it downward, and fell out of the tree, dragging all the ropes, the food bag, and a thicket of broken branches to the ground. My brilliant problem-solving came to naught; the bear consumed my food.

Both hierarchical reinforcement learning and model-based learning work well when there are just a few actions to plan as a sequence. But what of problems whose solution requires many steps? As steps are added, the number of potential solutions grows exponentially. Here is the curse of dimensionality in problem-solving in all its glory. Suppose you are bagging six different food items into two grocery bags. There are 7,200 possible ways to put three different items into each of the two bags. If you add the basic rules of grocery bagging (heavy items on the bottom, fragile items on the top, cold items together), more than 7,000 of the possible solutions are incorrect. It would take forever to solve this problem through hierarchical reinforcement learning. Similarly, there is no way that a person could, in a reasonable amount of time, think through all 7,200 possible bagging solutions to find the best one. Nevertheless, bagging at the grocery store seems fantastically easy. At international grocery bagger competitions, speed baggers can correctly load thirty-eight items into three bags in less than a minute. How is this possible? There is growing evidence that the brain can tackle these increasingly complex planning problems via a hybrid solution that combines the best of both worlds, model-free and model-based learning.[10]

To get a sense of how this works, consider the task of preparing breakfast for four people, the goal being to have all the dishes ready at the same time. Just as in grocery bagging, there are an overwhelming number of possible solutions, and a solution based on trial and error is impractical. Even if one tried to simulate in advance the best possible way to prepare the meal by enacting all possible sequences, it would take far too long to find the best approach because of the curse of dimensionality. The only way to succeed in making breakfast for four is to break it down into a hierarchy of subtasks like frying the bacon, cooking the eggs, toasting the bread, setting the table, and so on.[11] Ideally, each of the subtasks is already well ingrained by previous model-free learning. Then the person is free to build a mental model that can balance the demands across the different subtasks. The cooking challenge has actually been tested in a virtual environment. Participants were also tested on other aspects of intelligence.[12] Not surprisingly, those with greater cognitive resources—working memory, reasoning power, and self-control—were better able to break the task up and coordinate the different subtasks so the meal was ready all at once.

The ability to form a hierarchical model to problem-solve has long been associated with the brain's prefrontal cortex. A clear example of this can be obtained from healthy people tasked with negotiating a complicated subway system while being scanned by fMRI.[13] Anyone lost in the London Underground will appreciate this task. Trying to find a route through the lines and stations is a strong activator of the prefrontal cortex. But those with more life experience, like the elderly, have more knowledge and habits to draw from and don't always have to rely on this strategic kind of fluid intelligence.[14] Not surprisingly, patients with damage to the frontal lobes lose this global strategic planning and the ability to coordinate subtasks.[15] Even in healthy people, this ability to organize a plan hierarchically can be undermined by transiently inhibiting the inferior frontal lobe through transcranial magnetic brain stimulation.[16]

As I stashed my food in the bear box, I found I was testing out a new design. The handles and levers on the box were gone. Instead, there was just a small keyhole. A stout key hung from a metal lanyard, bolted to the side of the box, making it almost impossible to lose but available to anyone. The bear would need to grasp the key, put it in the lock, turn it clockwise, and pull the door open. This solution requires enormous imagination, or a long season of trial and error. The designer was assuming the bear lacked sufficient cognitive power for model-based reasoning and lacked the time to find a solution based on trial and error. Working the key into the hole, I had my doubts. The lanyard was frayed, the key was bent from bite marks, and the keyhole was surrounded by scratches. Perhaps insight was just around the corner?

8

Purpose

Take an object. Do something to it.
Do something else to it.

—JASPER JOHNS[1]

I NEEDED TO CROSS a particularly high pass because it was spectacular for its rugged topography and, more important, offered a huge shortcut. As I approached it, I felt escalating dread. The far side of the pass, the side I would have to descend, would make anyone nervous. It was a wickedly steep snow slope, a half mile down, and there was no way around it. Early in the summer it would be covered in snow, making it simple to trudge down by kicking nice deep bootprints into the snow. But it was late in the season, and the snow had almost certainly melted and refrozen countless times, making it slick ice, a nearly vertical glacier. Kicking steps would be impossible. It would be like trying to ski down an icy double black diamond run without the skis. There was no room for error. One slip and I would go the distance, bouncing all the way to the bottom, ending in a field of jagged boulders. I had no means to arrest myself if I fell. As I headed toward the pass, lost in my thoughts of all the ways I could fall down the icy run, I clumsily tripped over a solution. There before me was an old tree, growing out of a crack in the granite with crooked limbs reaching upward. One long limb, close to the ground, was particularly gnarled, worn bare from many a hard winter. Here was a tool I could use.

Prying off the dead limb, I dwelled on the quick transformation of this object. I instantly envisioned a way to use this tool, which was suddenly endowed with purpose. It made me wonder, Is purpose all that is required for something to become a tool? Or did the stick offer more? If only purpose was needed, then lots of objects are tools and lots of creatures are tool users, building nests, shelters, and burrows out of all kinds of material. So it would be that even earthworms could be considered clever tool users. Darwin, in his popular book on earthworms, described how they will pull a shred of a leaf over their wormhole when they sense an impending storm.[2] Earthworms manipulate the leaf to improve their comfort. Similarly, apes will use broad leaves to keep their heads dry when it rains. While endearing, the notion of leaf as tool, whether used by worm or ape, seems too loose. There must be other contingencies besides purpose for something to be considered a tool. What the leaf lacks is the property of a utensil, an instrument or implement that can be held in the hand, beak, mouth, or trunk, allowing the body to perform some kind of work that is otherwise difficult or impossible. It is this marriage of manipulation and purpose that gets us close to the notion of a tool. While my dog can carry a stick in her mouth, that is all she can do with it. With a stick in my hand, I could find a way down the ice field.

For over a century, a central dogma of human uniqueness was our capacity to use tools. Fortunately, there are sufficient examples from Animal Planet shows and other documentaries to finally kill off this antiquated idea. Instead, what has emerged is a far more interesting question about the different ways that animals use tools.[3] Their techniques are nuanced, but this variation concentrates the mind on what makes an object a tool. Some necessities can readily be discovered by a stroll through a hardware store. There are countless kinds of objects, requiring endless demands on physical skill and, just as important, insight into how things work. Hardware stores arrange tools in different aisles, depending on the problems they solve. Within an aisle they are further sorted by shapes or how each is handled. However, it is also possible to sort these tools another way: according to what the tool user needs to know about the physical world in order to employ a particular tool. Some of these concepts are extremely primitive, abilities that might be present in a bird brain; others are quite advanced, present only in the mind of a few aerospace engineers. The key insight from the hardware store is that there is neither a single physical attribute nor a single notion of what a tool is within the mind. There are gradations of tools and tool use that reflect fascinating conceptual abilities. So it was that I could envision the stick I had just picked up as four very different tools for descending the ice slope, each requiring distinct physical and conceptual abilities. My simple stick, now a multipurpose climbing tool, offered a window into a few of these primitive concepts and their underlying brain mechanisms.

One way I could use my stick was as an alpenstock. The alpenstock is a stout staff, like a very long broomstick with an iron spike on the end. Shepherds began using them in the Middle Ages as a means for herding their flocks across relatively flat, albeit slippery, glaciers in the Alps.[4] More than just a means for poking sheep, it could be used as a cane or walking stick to provide the stability of a tripod stance. The stick could also be used to poke around in the snow to feel for hidden crevasses. It could also be jammed down into the snow, forming a solid flagpole to hang on to as an anchor on a slippery slope. It remained the primary tool for foot travel through the Alps well into the 1870s. It was common for tourists to lug them home as souvenirs of their mountain visit. The primary means by which the tool worked was simple: it reshaped the human body so that the arm could reach much farther than normal, either to the ground ahead or into the deep snow. The concept that an animal requires to employ this type of tool is readily determined by the tool's shape: a direct elongation of an arm. This may seem self-evident, but, as we will see, there are many tools where the means to use them are not this obvious.

Most animal species that use tools hold a simple object so that their body can be reshaped. In the evolution of the physically oriented mind, this is a remarkable breakthrough. Animals discovered that by holding a stick they could fit into affordances that were otherwise inaccessible. A bird or chimp will use a stick to probe a nest to catch more ants or termites. Holding the stick, the bird ends up with a longer beak and the chimp with a longer finger. The beauty of this kind of a tool is that the brain can readily take advantage of the inherent malleability of the body schema. Computationally, it is just the elongation of a body map that is already being adjusted all the time as the animal moves.

While the idea of reshaping is theoretically a good one, demonstrating in the lab that a human brain actually makes this kind of adjustment is another matter, requiring a bit of ingenuity. In one clever experiment, Lucilla Cardinali gave people some grabbers, like the gadgets people use to pick up trash without having to bend over, and had them reach out across a table to pick up objects.[5] The tool naturally extends a person's reach to the far side of a table. The question was whether brief use of this tool would change the body schema. To measure the schema, the participants were next blindfolded, and the experimenter touched them on the arm that had used the tool, either at the elbow, wrist, or tip of the middle finger. The participants had to point with the index finger of the other hand to a spot directly above where they thought the experimenter had touched them, without actually touching themselves. The amazing result was that after the participants had used the tool for just a little while, the distance between where they felt their elbow and where they felt their middle finger were located had increased. They had stretched the scheme of their arm simply by extending their reach.

The incorporation of a stick or other simple object into the shape of a body doesn't seem far-fetched. All it takes is for the body schema to gobble up the stick itself, to make it part of the animal's body shape. In the 1990s, Atsushi Iriki at the Tokyo Medical and Dental University got a glimpse of how neurons in the brain might actually do this gobbling. He was studying how Japanese macaque monkeys learn to use tools. There are many noteworthy photographs of these monkeys crowded into remote hot springs, surrounded by the snowy mountains of northern Japan. While highly social, they rarely exhibit tool-use behavior in the wild. But in the lab they can be trained to use simple rakes, like the toy rakes kids use in a sandbox. After two weeks of training they could use a rake to draw in food that was otherwise impossible to reach. Iriki then recorded neurons in the parietal lobe of his trained monkeys, in the area that is associated with creation of a body schema. These neurons respond when the hand is positioned in a particular location in space, reflecting a location of the body schema. After learning to use the tool, a monkey would fire neurons for stimuli near the tip of the rake. That is, the amount of space where the neuron fired had expanded, as if the monkey had incorporated the tool into its body schema.[6] Reshaping of the body schema for using objects like sticks was probably one of the first ways that early hominins used tools. A small, thin stick could be used to nab hard-to-reach food like insects or honey or to shove marrow out of a carcass bone. A big stick could be used as an extra leg to provide increased stability in rough terrain. The cognitive machinery necessary to do this is fairly rudimentary: the shape of the object defines how the body schema can be reshaped to fit into affordances of the environment.

Cardinali and Iriki's experiments suggest that a stick or rake can become part of the body, as if the tip of the finger were transplanted out to the tip of the rake.[7] It turns out that where the finger gets "placed" along the rake is quite specific. It isn't placed necessarily at the distant tip, but rather at the position on the rake that needs to be controlled. To show this, another experiment with people as participants used two different rakes, one thirty centimeters long and another sixty centimeters, but with its tines at the middle of the shaft, thirty centimeters away from the hand. The second rake was twice as long, but in terms of functional use it was identical and induced the same amount of expansion of peripersonal space.[8] From an engineering perspective, this kind of adjustment is exactly what you want the brain to do: to plan movement in terms of a body map centered at the exact spot that will interact with the world. In this experiment, it was wherever the rake tines ended up.

These careful adjustments of the body schema to incorporate an object raise another important conceptual question. When the stick or rake is gobbled up by the body schema, does the tool become part of the monkey or person's identity? That is, can the animal reap the benefits of having an elongated body schema with its new control point and at the same time have a concept that distinguishes the self from the tool? When a North American grackle uses a stick to retrieve food, it may not be able to distinguish where its beak ends and the stick begins. Conceptually it might perceive that it has a longer beak. It is obvious that people don't confuse a stick with ourselves. We will use a long stick to poke a log in the fire with no worry that we will get burned just because the tip of the stick is in the coals. This self/tool distinction was almost certainly in place in our hominin predecessors. We can be sure of this because of an amazing archaeological discovery in Schöningen, Germany. There, eight wooden spears dating from approximately 350,000 years ago were discovered, located among the bones of slaughtered horses, with one of the spears sticking out of a horse pelvis.[9] Much has been made of the fact that to do this kind of hunting of a faster, bigger animal, the spear users must have worked in teams. From the perspective of tool use, the spears also provided a way to hunt the horses while maintaining safe distance. Any confusion over what was person and what was spear would have made this kind of hunting impossible.

Do other species have this concept of tool as separate from self? For macaque monkeys, whose tool use in the wild is questionable, we really don't know. Chimpanzees, on the other hand, are adept tool users and give us an opportunity to test for this kind of distinction, as Dan Povinelli and Scott Frey did in a brilliant experiment with chimps conducted at a colony in Louisiana.[10] Their chimps were offered a box with a much-desired apple in it. Not surprisingly, they would reach with their hand to nab it. In some trials, the box contained an aversive object like a rubber snake that the chimps would naturally avoid. If a chimp saw a snake, it wouldn't put its hand anywhere near the box. The chimp was next offered a box with cloth over it. Did it have an apple or a snake in it? Pulling the cloth off with its hand was too risky for the chimp, which was provided with a stick. If the chimp used the stick, it would show that the chimp was willing to risk the snake biting the stick. That is, it would know the difference between tool and self. Not using the stick suggested that the chimp's body schema and body identity extended to the tip of the stick. The results were clear: the chimps used the stick to remove the cloth. They could distinguish between tool and self. A simple experiment like this reveals how a single conceptual advancement, such as the ability to distinguish between self and an object, can dramatically expand the ways in which a tool can be used. A stick becomes a lethal hunting tool for hominins and a means of avoiding injury for the chimp. Given that, it is interesting to speculate why macaque monkeys are not particularly great tool users despite having excellent hand dexterity. Perhaps their limits as tool users derive in part from their inability to know the difference between self and tool, greatly diminishing situations allowing for the use of a tool. If so, then what Iriki might have been doing by training his monkeys to use a rake was less about tool use and more about altering body schema.

As I headed for the ice, I already knew that a tool like an alpenstock wouldn't suffice. The slope was too steep and too hard. If I started sliding I wouldn't have enough time to jam the tip of the stick into the hard surface. I needed another kind of tool. By the late 1700s, alpine explorers were doing much more than just herding sheep. They were climbing the local peaks, and, to access the really tall ones, the more aggressive climbers had to find ways to ascend steeper ice. They brought along a second tool, a hand axe. When Jacques Balmat and Michel-Gabriel Paccard made the first ascent of Mont Blanc on August 8, 1786, Balmat had his trusty alpenstock as well as a hand axe, the kind you would normally use for chopping wood. Balmat used his axe to chop steps into the ice to fashion a rough-hewn ladder. By the late 1800s, climbers had discovered that a hand axe was far more effective if they rotated the blade 90 degrees relative to the handle, inventing what would become known as an adze. They also shortened the alpenstock to the length of a cane, kept the pointed metal tip on the bottom, and stuck the adze onto the top. This was the forebear of the modern ice axe. The adze, combined with the short shaft, made it easier to cut steps both above or below the climber's stance. Balmat's axe and its twisted cousin the adze are canonical tools, direct descendants of the earliest found tools used by hominins and not all that different from the copper axe Ötzi was carrying in the Alps five thousand years ago. The archaeological record describes a lineage of metal axes, preceded by stone axes, preceded by carefully honed sharp-edged rocks, which were almost certainly preceded by blunt rocks. This lineage reveals the emergence of another fundamental concept that introduced an entirely new form of tool, a concept predicated on an understanding of dynamic force. Many animals can apply powerful forces with their beaks, jaws, claws, or horns. The conceptual leap for tool use was understanding how to focus and amplify energy not only with the body, but through the medium of a held object.

To appreciate the importance of this concept, spend some time at play with a three-year-old. When challenged to crack a tough nut, he will try all sorts of ineffective methods, revealing a very general knowledge of physical forces but nothing particularly useful about the force-focusing or -amplifying capacity of tools. He will use gravity as a force generator by dropping the nut on the ground or a rock on the nut. Or he will try using forces generated by only the body, such as throwing the nut against a wall or hitting it. Or he might try banging the nut with a rock or hammer. All instances of indiscriminate exploration. By the time the three-year-old is five, he will have discovered the relationships between force, mass, acceleration, and the effective deposition of energy via a handheld tool. By then, the child will have intuitive knowledge of the fundamental laws of dynamics (without ever taking high-school physics) and can generate a precise swing of a five-pound rock, effectively doubling its kinetic energy. By eight, he will also be able to use an anvil to pound the nut, revealing an even deeper understanding of how energy can be concentrated into an object.[11]

The ability to conceptualize dynamic force is rare in other species. Like five-year-olds, chimpanzees in the wild have some semblance of it. They will use a stone and anvil to crack open a tough nut. They clearly have an understanding of the utility of concentrating force. And they will pick out a suitably weighted rock for the task. One can look harder and test how well they understand the dynamic aspects of the task. Do they simply steer the rock downward and rely on gravity to generate the force, or do they know something deeper about the physics of the problem? To test this, Blandine Bril and Julie Foucart provided captive chimps with stones of increasing mass to see how well they adjusted their arm swing to control the forces applied to a nut.[12] Just as important, by looking closely at a chimp's movements, they could determine how much "heat" the chimp added to the weight of the falling stone to give it more cracking energy. Increasing the velocity of the falling rock beyond what gravity created would show that the chimps had a very sophisticated concept of dynamic force. The videos of their one-armed thumps are rather sad. Their form is clumsy, as if they could use a lot more practice swinging rocks. Nevertheless, across a range of stones, they added a modest amount of oomph to the downward blow of the rock swing and made careful adjustments that depended on the mass of the held rock. Clearly, they have a fairly solid concept of dynamics as applied to a tool. But one walks away underwhelmed with their competence.

To watch the king of nutcrackers, you need to go to Brazil. Deep in the forest there are many New World monkey species like the bearded capuchin monkey who are spectacular rock pounders. Capuchins are those adorably clever little monkeys, familiar as pet organ grinders or rascals in old-time movies. In the wild, a large part of their diet comes from the tucum nuts of the palm tree. Like Brazil nuts, these are really hard to crack with bare hands or teeth. Given the demands of their ecological niche, it is no surprise that the capuchins have mastered the use of stones as pounding tools. Madhur Mangalam and Dorothy Fragaszy of the University of Georgia wanted to quantify the capuchin's level of skill, so they set up a video camera in the wild and recorded fourteen different monkeys at work.[13] A monkey will place a nut on a large base rock, which serves as an anvil. Then the capuchin will use both hands to lift a stone weighing roughly half of its body weight above its head, in a continuous smooth motion like the clean and jerk of an Olympic weight lifter. Using its entire body, it rams the stone downward like a pile driver, augmenting the acceleration provided by gravity to smash the nut. There is no doubt, watching them at work, that capuchins are adding an enormous amount of energy to the stone strike. They clearly have a concept of dynamic force mediated through the stone.

Careful video analysis reveals that there is more to their skill than simple power lifting and pile driving. Too strong a hit and the meat of the nut becomes mush, making it hard to separate from the shell fragments. The best way to crack the nut is with a couple of hits, each with less than maximal force. Mangalam and Fragaszy could show that the capuchin's whole-body pile drive required an incredibly refined level of dexterity, finesse, and efficiency to generate just the right amount of force. Comparing the poor performance of juveniles with the precision of adults, it was apparent that competence takes years of practice. While young capuchins could imitate the adults, it requires physical practice to become a consistent nutcracker. Furthermore, the adults are picky about which rock they use.

They understand that dynamic force involves both the forces applied to an object and something about the object itself.

Different ecological niches likely drive the competency of mastering nut cracking: an essential skill for a capuchin and a marginal one for the chimpanzee. To leverage the mass of a rock as a tool, the capuchin is transcending an understanding of how things move in the world; it is learning something profound about the causal properties of movement. A nut falling from a tree and hitting it on the head is just an event; there is no causal explanation for why it happened. That is the way much of life is experienced by animals. In contrast, holding a rock in the hands to smash a nut makes sense only if there is an understanding about the relationship between something the rock has to offer (mass) and a desired outcome. This may seem obvious, but understanding objects of the world in terms of their dynamic rather than kinematic properties—what can be done through them rather than what they do on their own—is an overwhelmingly profound conceptual leap.

Being able to lift a rock with one hand requires an enormous amount of visual and motor coordination. At a minimum, there needs to be an understanding of an object's 3-D shape, its probable mass distribution, and the appropriate hand shape to match the object. It is a two-way integration of seeing what is important about the object and knowing what the hand can do. The brain circuit that allows a chimp or human to pick up an object, whether a large rock or pebble, was identified in the 1990s by an international team that I was lucky enough to work with. One lab, led by Hideo Sakata[14] in Tokyo, Japan, recorded neurons in a portion of the parietal lobe of monkeys that he referred to as "area AIP." This area is positioned adjacent to regions that are critical for touch sensation. Sakata found that neurons in the AIP area would be more active when the monkey looked at one object in particular compared to another. Each neuron preferred a particular 3-D shape or object geometry. It might fire the strongest in response to a cylinder, or a cone, or a cube. In a critical manipulation, Sakata turned the lights out after the monkey had seen the object, so it had to grasp it in the dark. Some of the AIP neurons also fired during the delay or when the monkey reached for the thing in the darkness. Thus the neurons represented information about the object and its physical properties independent of the visual stimulus. Inhibiting this perceptual area with a drug knocked out the monkey's ability to shape its hand to match an object's shape. In later work, Sakata showed how the AIP area also decoded the three-dimensional properties of the object, invaluable information that their brain could use to determine how to pick something up.

In parallel work, Giacomo Rizzolatti led a team in Parma, Italy, that recorded neuronal activity in the ventral premotor area (PMv) and neighboring cortical regions of the frontal lobe in monkeys.[15] Rizzolatti trained monkeys to reach for and grasp objects such as pellets of food. The animals' AIP and PMv neurons clearly fired in relationship to specific hand actions determined by specific objects. Anatomic studies from Massimo Matelli and Giuseppe Luppino of Parma revealed rich interconnections between AIP and PMv.[16] In 1992 they were joined by Marc Jeannerod, a cognitive neuropsychologist in Lyon, France. Jeannerod provided extensive behavioral evidence to support this two-module architecture. He had studied the timing of natural reaching and grasping movements in people for decades and knew that people who had sustained a stroke affecting the human analogue of area AIP did not shape the hand correctly when grasping for an object. Michael Arbib of Los Angeles, a computer scientist, had built models to train robots to manipulate objects that were reframed in relationship to what the brain was doing. As part of the AIP-PMv team, I used brain scans to show where this network was located in healthy humans.[17] Our group proposed an entirely new way of thinking about how the brain accomplishes something as mundane, but ultimately as complex, as object grasping.[18] Since then, many other scientists have added to this model. For example, it is possible to track how the arching of the palm is also exquisitely timed during the reach of the arm so that the entire shape of the hand, not just the distance between the fingertips, is configured to approach and touch the object at just the right moment.[19]

The most interesting computation performed by the grasp circuit is matchmaking between an object and the hand. To be a good match, it needs to be reciprocal: Given an object, what is the best shape for the hand? Just as important, given the possible shapes of the hand, what objects will fit it? This linking of perception and action into a common framework is clearly a learning process that requires bountiful physical experience. It starts early in development; at eight months of age, human infants are already beginning to use a precision grip. Once they can do that, they are more likely to be able to recognize goal-oriented precision grips performed by others.[20] By adulthood, the human AIP-PMv circuit is a powerful matchmaking engine, linking objects in the space near a person with possible actions with those objects.

One of the first studies to show this automatic action planning was a brain-scanning experiment I performed with Giacomo Rizzolatti. We prepared for the experiment by ransacking my house, filling a large box with every graspable item we could find. At the lab, we had participants in a brain scanner looking, grasping, and naming each object. One of the surprising results was that compared to a control scan where a participant just looked at an ungraspable fractal image, we found that the mere sight of a familiar object activated premotor areas known to be involved in movement planning.[21] We also discovered that when an object is nearby, it automatically and rapidly grabs a bit of our attention, even when the object is completely irrelevant.[22] For example, having a saltshaker within reach on the dining-room table will grab a little bit of your attention, even if you don't want to salt your dinner. This automatic engagement of the grasp circuit is fast, within 140 milliseconds of the appearance of an object, and it takes only 250 milliseconds to recognize how to use something.[23] Recent evidence suggests that much of the automatic activity in these kinds of experiments is dumb planning, the fingers and thumb seeking positions that will encase what the observer thinks is the center of mass—the most stable position for lifting but not necessarily the grasp needed to use the object as a tool. To do something, a simple grasp needs to be augmented by a deeper understanding of what an object can do functionally.

The application of dynamic force, like pounding a nut with a stone, has probably been evolving in one primate species or another for many millions of years. In light of that, do capuchins have any understanding or intuition of the pounding rock's weight? This may seem like an inane question, since they can clearly learn to pick out the correct rock to use, and once they start swinging, they are quite good at controlling the smash. But it's possible that they might learn to pick out a rock based on size rather than heft and learn to control the pounding without really knowing anything about weight. The question is whether they have an ability to make a mental model of weight without actually picking up an object. If you look at a bowling ball, it is easy to have an intuition that it is heavy, and your body posture will anticipate this weight as you get ready to lift it. This kind of inherent knowledge and anticipation can be measured formally in the lab by assessing the amount of force a person applies to the fingertips to grip and lift an object.[24] The more one knows about an object, the better the anticipatory grip or lift forces, and once manipulation starts, fewer corrections are needed to maintain control.[25] In an exhaustive series of experiments, Dan Povinelli showed that chimps don't have the kind of anticipatory knowledge that would help adjust the appropriate forces.[26] All the chimps seem to do is react with a barrage of quick adjustments as they manipulate an object. This is in part why their interactions with tools always look so awkward and clumsy. Without an intuition about or knowledge of weight, they can only react to what the object does to them. It can be argued that this lack of insight into weight places a harsh limitation on what they can fabricate.

To focus energy beyond what a round stone can deliver, a sharp edge is needed. Recognizing the value of a particular shape, like a point or a sharp edge, is in itself fairly common across many different species of tool users. The more dramatic conceptual leap most certainly would have been the emergence of fabrication, the ability to create a sharp edge from a round stone. Granted, birds can fabricate nests, but these are static objects where building materials are sequentially added. To fabricate a stone tool requires much more. While the earliest stone tools of our hominin ancestors look brutish, their creation (called knapping) is unbelievably difficult. It requires the maker to bang one rock against another, with extremely strong blows applied at just the proper angles. The effort was worth it. The tool's sharp edge, backed by a lot of mass, made impossible tasks easy. The maker could crack bones to get their marrow, split nuts for their meat, and cut down prey. Knapping required the maker to have an intuition of the weight, density, texture, and center of mass of the stone in anticipation of how he might chip out an edge. This insight was already present in whoever made the oldest-known stone tools 3.3 million years ago.[27] The concept of weight is deeply ingrained in the modern human brain's perceptual system. Weight and density are decoded in visual areas including the ventral occipital cortex, a site that is primarily involved in object recognition.[28] When you view a coffee cup, you are doing more than just finding a way to name it. You are also retrieving knowledge about its hidden physical properties, like expected weight or density. The part of the brain involved is strongly connected to the human homologue of the AIP area, the region used to process grasping. This connection binds together two different visual systems, creating a powerful mechanism for linking conceptual knowledge about object properties like weight or center of mass with the possible ways to grasp an object.

Knapping demands powerful grip strength to hold each of the rocks as they are clacked together. Capuchins don't have this kind of power; the best they can do is use two hands to lift a rock overhead. Chimps have a weak grip aggravated by a short thumb. When can a hand that is strong enough to make a stone tool be found in the fossil record? Around three million years ago, most of the existing fossils are of a short, upright-walking primate called *Australopithecus afarensis*. This species is famous for a nearly complete skeleton that was named Lucy. It is possible to compare her bones to those of other living species in terms of length, robustness, curve, bony bumps where tendons attach, and the way adjacent bones slide against each other. Every new fossil discovery from these oldest bones hints at features consistent with modern hand functions like precision or power grasping or, just as likely, toward nonhuman functions like tree climbing.[29] Lucy and her relatives have a complex mixture of hand features that largely suggest functions found in tree-climbing primates. But they also reveal something new. The hand of *A. afarensis* has a relatively long thumb compared to that of chimpanzees or gorillas. There is also a bony protuberance on the thumb, which means there was a tendon attached to it that strengthened the thumb's flexing power. The fingers are thinner and more delicate than those of a full-time tree climber. Because this is the only species found from this time period, it is tempting to fill in the gaps and conclude that they made tools. In fact, there are experts who express doubt about *A. afarensis* as a toolmaker. Stone tools have never been found alongside fossils from Lucy or her species, and there is currently a consensus that she was probably a creature living in and out of trees with modest hand function. For all we know, there could have been a completely different, unknown species alongside Lucy that was making those first tools.

Jumping forward a million years, the fossil record begins to pick up. By two million years ago, there were at least two new and very different primate species present, together with an abundance of stone tools.[30] The first species is called *Australopithecus africanus,* of which there are enough bones and skulls to recognize a number of humanoid features. As for the hands, the length of the thumb compared to the fingers is similar to modern humans. The bones of the palm are thick, allowing for a much stronger grip. There are remnants of tree-climbing features in the fingers, but they are becoming less prominent. The other recently discovered species is called *Australopithecus sediba.*[31] It has a mash-up of bony features unlike anything else. The thumb is long, really long. Imagine if your thumb was an inch longer, reaching almost to the length of your index finger. The bones in the palm of *A. sediba* are quite delicate, limiting its grip strength, while the wrist bones, particularly at the base of the thumb, are quite stout, allowing for a strong grip. This combination must have provided a very different kind of opposing grip, but for what purpose? This was a hand adapted to an environment that we don't yet recognize. The contrast between these two species shows that by two million years ago, at least a couple of species had sufficient cognitive insight and hand morphology not only to apply dynamic forces, but also to create tools. There would be selective advantage for any species that could do the rough work of knapping.

I finally reached the pass. It was bracketed with serrated ridges rising on either side, blocking any alternate routes. Beyond the pass, my walking ended abruptly. Before me was a massive ice field, dropping vertiginously a half mile downward. The ice was coarse and crunchy, far too hard to kick a step into. In theory, I could have chopped my way down, using my stick if it had a branch at the end shaped like an adze, similar to tools that have been around for at least a million years. However, chopping out a half-mile staircase would have been a horrendous chore, and a wooden adze wouldn't sustain that amount of chopping. And if I fell at any point, I might easily slide out of control. I needed a way to grab the ice. To solve this problem, alpine climbers added a third implement to the ice axe in the 1870s—a pick, a knifelike blade sticking out at a right angle from the top of the wooden shaft. By holding the shaft, the pick could be hammered into the ice, boring in like an enormous talon. More than just a means for elongating the body, as the alpenstock did, or a way to transfer energy, as the adze did, the pick of an ice axe transformed the climber's hand. The pick amplified a clawlike grasp already present in the hand.

To get down the ice field, I could use my stick with its pick-shaped branch at the end like the pick of an ice axe. This would be much easier if I had a tool for each hand—then I could work my way down taking turns hammering the picks into the ice. That would be a lot of work, but more worrisome, if I fell I wouldn't be able to hammer in the pick and hang on to it if I started sliding too quickly. I knew this only too well. Years before, I witnessed a climber make this mistake while we were climbing one of the volcanoes near Mexico City. Although one thinks of Mexico as hot and flat, just outside the city there are two enormous glacier-topped volcanoes. Partway up one of the glaciers, there is an ice slope that starts out gradually, bottoming out in a flat area. I watched a friend fall out of control on this section, ending in a pile of soft snow and a lot of laughs. It was as fun as a sledding hill. Higher up, the ice field became steeper and angled to the side, until it hung above a six-hundred-foot cliff. On another trip, I watched a climber on this upper section slip and fall headfirst down the slope. He had a fixed concept about how to use his ice axe. He swung the pick at the ice like a hammer, where it stuck fast, but he couldn't hang on to it. Within moments he was hurtling down on his back at more than forty miles an hour. He froze in a crumpled shape, not knowing what to do and unable to stop. Time seemed suspended as he plunged off the cliff, launched into space, and landed out of sight far below. After coordinating his retrieval, I went up and finished the climb, walking in his footsteps. At the spot where he fell, all I could think was how senseless his mistake had been. It was a steep section but entirely manageable. Knowing just a little bit more about how to use his ice axe, he could have readily stopped himself. He needed a more flexible concept of how to use his ice axe besides hammering or grabbing.

The idea that tools and actions are united in the mind as flexible concepts became apparent in 1900, due in large part to the groundbreaking insight of Hugo Liepmann, a Viennese neurologist.[32] He had a patient who suffered a stroke, rendering him unable to speak. Nevertheless, the patient, a forty-eight-year-old imperial councilor, could follow instructions such as "Walk to the window" or "Stand up." Liepmann noticed that the patient performed tasks like buttoning his shirt haltingly. He wasn't paralyzed—he could button a button. But on completing one button, he had trouble moving on to the next. Similarly, he had trouble putting all the steps together that are needed to light a cigar. Liepmann asked the patient to show him how to play an imaginary harmonica, to which the patient made some awkward gestures of the hand. Liepmann named the deficit "ideomotor apraxia," a disorder that has now been studied in detail for more than a century. The hallmark finding is the clumsiness of complex actions involving tools, particularly when the patient has to recall the action from memory to pantomime it with an imaginary tool. We now know that a lesion of the supramarginal gyrus, a section of the parietal lobe, or a reciprocally connected region of the frontal motor areas will result in this deficit.[33] In my lab we found that healthy participants who were asked to pantomime how to use imaginary objects, like a hammer or scissors, activated these same brain areas.[34]

In addition to reflecting what a brain knows about tools and how they are manipulated, patients with apraxia also reveal something new and exciting about how tools fit into the broad domain of physical intelligence. Any object can become a tool once it is associated with an action and a problem that can be solved. It is the marriage of function and problem that makes a tool. This conceptual level of thought can be tested by asking apraxia patients to match pictures of different tools (a pipe wrench and a hammer) with different scenarios (a leaky faucet and a board with a nail in it). Here too, brain imaging experiments reveal separable brain systems that represent physical knowledge, closely aligned with tool knowledge.

It has often been argued that this kind of conceptual insight must have appeared with the emergence of language. It is hard to tell whether an experiment where the patient is told what to do is really testing something about action concepts, since the experiment depends heavily on language. Instead, the question has been reframed: Is action knowledge supported by a dedicated set of brain areas? Is action a distinct kind of knowledge? By using brain scans of healthy people and studies of patients who have had a stroke, the relationships among concepts of actions, tools, and language can be unpacked. From scans, it is clear that brain areas that represent ideas about actions and tools are in places different from those for faces, cars, houses, environments, and so on. Actions and the use of tools constitute their own special category of knowledge.

It was time to get down the slope. My trusty stick with all of its multipurpose tools did more than make the descent possible, it made it fun. I had two tricks up my sleeve—tricks that came from prior experiences and a very flexible concept about how to find a function of the stick that would solve my immediate problem. The first trick was a glissade. I sat on my rear and slid down the slope, holding the stick across my lap, pushing my weight down onto the trailing tip dragging behind me, effectively making a combination brake and rudder. When done properly, it turns an ice slope into a half-mile playground slide. The second trick was self-arrest. If I got going too fast, I could always apply full power brakes by rolling over on my abdomen and holding the stick underneath me with all my weight on top, just like I would do with an ice axe. The stick had a hooked tip, like the pick of an ice axe, that I could drive into the ice just below my sternum. Neither of these maneuvers is hard to do, but they are counterintuitive movements, not the first actions a person would be tempted to perform. They are hidden functions of the tool, functions that were invented and reinvented by climbers over a century, all hidden inside a piece of wood. Down I went.

9

Costs

The journey must now, I presume, be conducted on foot; indeed, this is clear, since we have no vessels; and we are driven to commence it at once, for we have no provisions if we stop.

—XENOPHON[1]

I CRAWLED OUT OF the tent and rubbed my eyes to get a look at the neighborhood, a maze of what should have been shimmering lakes and looming peaks in the dawn light. But the shimmer was gone. The vision in my right eye wouldn't clear. It was as if someone were holding a veil in front of me, like a crumpled screen door or a lace handkerchief. I rubbed my eye some more, hoping it would go away. Looking carefully, it was worse than I thought. There were floaters, countless little dots like black flies swirling all across the right eye's visual field. They weren't flies. They were blood cells floating around inside my eyeball. Looking to the left or right triggered bright flashes, like lightning bolts, a sign of unstable neurons in the retina. The motion of the eye was juggling them so groups would discharge in volleys, creating illusory images. The veil kept flowing about, blocking the view, a sign that the vitreous humor, the jelly that fills the eye, had pulled away from the retina. All these problems can occur with eye trauma, but more commonly they just pop out of nowhere, signs of an old eyeball.

It can be a not-so-bad problem that goes away on its own, or it can be a catastrophe if the retina gets torn, culminating in permanent visual loss. I knew all this because I had already been through the calamity with my other eye, which had never healed. With one eye fuzzy and the other eye hazy, I was nervous that my failing vision could get me lost or worse. My adventuring was over. Getting to a specialist out in the "real world" was paramount. I packed up fast, trying to think positive thoughts and getting myself psyched up for what would be a very full day. It would be a twenty-four-mile walk, mostly cross-country and over three passes, about the same effort as walking thirty miles on flat ground. I headed off, picking the easiest possible route, walking as smoothly and rhythmically as possible, taking long strides and using full arm swings, augmenting my uphill step with sporadic shoves from a pair of skinny walking poles. Even though the world was a foggy blur, I felt good. If there is one action that humans are designed to do well, it is walking.

Many paleoanthropologists have made the case that walking efficiency is a particularly important outcome of relentless evolutionary pressures that shaped the physical design of the hominin body.[2] Endowed with endurance, our ancestors had the ability to travel exceptional distances to find food, whether it was to scavenge for a carcass downed by other predators or to endlessly stalk a less resolute cervine beast until it became exhausted first. Either trek would have benefited from a biomechanical design minimizing caloric consumption. Human biomechanics are most efficient at about three miles an hour, and the current record holder kept up this speed for more than thirty days. However, that record was achieved with snacks and support from friends and family along the way. What eventually halted the record holder in power walking was a need for sleep. A more compelling measure would be endurance without any support whatsoever along the way, the situation our ancestors were more likely to encounter. How far could you go if you just walked out your front door and kept on going without food or water? There are many epic tales of ordinary people taking this kind of walkabout. I'm inspired by the autobiographical narrative of the French pilot and best-selling author Antoine de Saint-Exupéry.[3] Just after the Christmas of 1935, he and his navigator Prévot were on a flight from France to Vietnam in their Simoun 1 aircraft, a sleek and particularly fast one-engine mail plane. After crossing into North Africa, they lost their way in dense clouds. Dipping down to get underneath them, they plowed unexpectedly into a sand dune somewhere in the vast Libyan desert. Over the next three days they walked 150 miles over a dry hardpan landscape with nothing but an orange, a little piece of cake, a cup of coffee, and a few sips of wine to keep themselves alive before they were finally saved by nomads. This is a perfect demonstration of what anyone might be able to do, push comes to shove. According to Saint-Exupéry's memoir, he and his navigator had no physical preparation. When they weren't flying, they were spending as much time as possible hanging out in southern France, drinking wine and loving life. They never planned, practiced, trained, rehearsed, or ever imagined they would have to walk as far as they did. Admittedly, by the last day they were delirious, suffering uncontrollable hallucinations from severe dehydration. Nevertheless, their three days on foot are a stunning manifestation of the pure physical intelligence inherent in us all.

We usually don't think of walking as an act requiring much in the way of intelligence. But in fact, walking, more than any other action a person does, demonstrates a beautiful calculus that the brain endlessly performs to inform movement with grace, stability, and, above all, efficiency. On paper, the math behind efficiency is really simple; it is an account of how cheaply, as measured in calories, an action can be performed. The body runs like a house furnace, burning oxygen and fuel to release energy. The less fuel used to heat a house, the better. While simple to calculate, efficiency can be dodgy to measure, particularly in a person. Most people get a rough sense of their fuel management by tracking their weight, to see if their fat stores are slowly climbing or (as many hope) declining over weeks to months. If weight is steady, then calorie counting all meals provides a measure of daily fuel intake and ultimately what a person will require; about two thousand kilocalories ("calories" in everyday usage) for a medium-sized man. Fitbits and similar gadgets that measure how fast the heart is beating can make a slightly more accurate guess at how many calories are burned while exercising. However, these too provide relatively sloppy estimates of efficiency. And all these estimates are irrelevant if a person isn't replacing calories burned, like Saint-Exupéry and Prévot as they toiled across the desert burning up their own muscles for fuel.

To get a precise measure of walking efficiency, a much better approach is to determine out how much oxygen the walker is consuming at any given moment. Just as heat and carbon dioxide are released when wood is combusted with oxygen, caloric energy to move muscles is generated by "combusting" sugars, fats, and proteins with oxygen (and water to help break down the fats). To test how much oxygen is consumed, the walker breathes through a face mask connected to a tube loaded with high-tech sensors that measure the difference in amount of oxygen going in and coming out of the lungs and the amount of carbon dioxide produced. There are other fancy methods that measure changes of trace amounts of exotic isotopes in the walker's blood, but the oxygen method is a tried-and-true approach. Using this device, countless laboratories have charted fuel consumptions at particular speeds, stride lengths, and inclines, with or without loads; the effort is not that dissimilar to the miles-per-gallon calculations provided by manufacturers of every kind of car and truck. The key measure is the cost of transport. Rather than the amount of gas it takes a car to go a mile it's the number of calories it takes to walk a mile.* For the average-sized person this is about fifty-five calories. To make this more concrete, calories can be converted to the amount of Snickers bars it takes to walk a mile (above and beyond the energy needed just to stay alive). For a medium-sized man, the added cost of walking a flat mile at an easy pace is a fifth of a Snickers bar, roughly one hefty bite.[4] That's a pretty cheap cost of transportation. While people around the globe vary enormously in how fat they are, their cost of transport is pretty much the same everywhere, whether a person is a hunter-gatherer, urban dweller, or farmer.[5] Despite these differences in size and shape and the way people spend their days, each brain has figured out how to maximize walking efficiency under an enormous range of operating conditions.

Getting to that high level of walking efficiency required each of us to practice for thousands of hours as we progressed from crawling to cruising to toddling and, finally, to proper striding.[6] Once an efficient gait is acquired, what more does the brain have to learn? Maybe nothing. If a person climbs onto a treadmill moving at a fixed speed, she will find her most efficient cadence and stride length in two to three steps.[7] Something inside her nervous system is already highly calibrated to walk extremely well under a broad range of speeds and inclines. However, as it turns out, people never stop learning to walk. The brain is constantly tuning and adjusting movement to find the best possible stride length and cadence to maximize efficiency in a kind of superfast learning. These adjustments can

*For a 2015 U.S. vehicle it is a tablespoon of gas.

be hard to notice since people are already extremely efficient walkers.

To capture quick learning in the lab, a trick has to be played on the walker. To do this, Max Donelan and his students at Simon Fraser University measured efficiency with the oxygen-consumption method in healthy participants who walked on a treadmill at a steady speed.[8] Not surprisingly, people settled into a walking pattern that minimized the cost of transport. The trick Donelan used was to strap an exoskeleton onto the participants' legs. An exoskeleton looks like a knee brace, only it has computer-controlled servomotors built into it that can add resistance to movement across the knee joint in ways that Donelan could adjust. He set up two scenarios. In one, the quicker a person's cadence (the faster the knee flexed or extended), the stronger the resistance at the knee. This would be like trying to walk through molasses. To be more efficient in this case, the participants should learn to use a slower-paced, longer stride. In the other scenario, the slower the cadence (and the slower the knee flexed or extended), the greater the resistance. To be efficient, participants should learn to make quick-paced, short steps. For both scenarios, once the participants had explored the effects of the exoskeleton at a variety of walking cadences, they were able to find their new optimum stride and rhythm in about five to ten minutes. Somehow, the brain was able to measure movement efficiency and minimize the cost of transportation incredibly quickly.

Once they were done learning, Donelan tested whether this was their new normal way of walking or just something weird participants had learned to accommodate their knee braces. He turned the exoskeleton off and measured how long it took the participants to get back to their original, regular, non-exoskeleton way of walking. Remarkably, this return wasn't particularly fast. Despite decades of walking without an exoskeleton, it took people a few minutes to adapt back to their natural way of walking. His experiment shows how the brain will constantly try to minimize the cost of walking if given the chance and will quickly settle into a new normal.

The never-ending adjustments to walking that Donelan measured in the laboratory can also be seen in natural settings. There is an ultramarathon race in the mountains to the east of Mont Blanc, high in the Swiss Alps, called the Tor des Géants that is unquestionably one of the hardest trail runs in the world. It covers 205 miles and involves 78,740 feet of vertical elevation change, equivalent to almost three trips up and down Mount Everest. The contestants in this event are supreme athletes, with years of training and experience. For the 2014 race, the researcher Gianluca Vernillo had the brilliant idea of testing if their cost of transportation would change over the course of the run by testing participants at the start and finish of the race using oxygen-consumption machinery and a treadmill.[9] One might assume that after completing this brutal course, which takes almost five and a half days, the runners would be utterly fatigued and their form would be a mess, severely elevating their cost of transport. We know that form matters: a slight limp will increase a person's cost of transportation by 5 to 10 percent.[10] Two twisted ankles can double the cost.[11] What Vernillo found was surprising. The racers all maintained good form. No limping, no ragged gaits. After five days of racing, their cost of transport had actually decreased, particularly for walking up a steep incline. Just as Donelan had shown, when efficiency matters in a particular action, the brain will do everything it can to minimize the cost of transport. This kind of learning happens whether we throw a backpack on or haul a bag of groceries. Having walked for days with a backpack, I was well tuned to move efficiently as I headed for home.

Whether in ultra-runners or exoskeleton-framed treadmill

walkers, the results show that the brain is constantly trying to find efficiencies. If efficiency is a fundamental property of taking action, then it might also shape other behaviors. In a laboratory at the University of Colorado, Alaa Ahmed and her students have been doing experiments like Donelan with participants who make reaching movements rather than walking.[12] Her participants learn to use their arm to push against a robot arm that pushes back, analogous to the way the exoskeleton adds resistance as a function of the speed of joint rotation. With practice, the pattern of resistance in the robot arm can be learned and a participant can make fairly smooth movements in a nearly straight line. The beauty of this experimental design is that it allows Ahmed to also measure how much a person uses muscle stiffness to work his arm against the robot. To go in a straight line against a complicated force field, a person could use a strategy of tightening up all his muscles so his arm becomes very stiff, just as a halfback in a football game can stiff-arm his way straight up the middle of the field. While this makes him very stable against perturbations, it also takes a lot of energy to plow through the defenders. That's okay for a halfback being jostled about but wasteful for everyday actions. If this stiffening were happening in her experiment, Ahmed and colleagues would expect to see oxygen consumption rise. However, what she found was the opposite: a 20 percent reduction in energy consumption as people learned to use fewer muscle activations in the arm to move through the force field—akin to an agile quarterback sneaking through the middle without bumping into anyone.

The first insight into where the brain might refine movements so that they are smooth, graceful, and efficient was provided by Gordon Holmes, who used his X-ray methods near the front lines in World War I France to study soldiers who had been shot in the back of the skull, just above the neck. In a classic paper, he described their movements as clumsy, indirect, and inefficient.[13] Walking, his patients looked like sailors on the deck of a rolling ship, staggering along with wide-based gaits. Their reaching movements took a zigzag rather than a direct path, and there were lots of extraneous body movements that did little to smooth things out. Quick motions would overshoot the mark. Eye movements sputtered. Holmes also manipulated their limbs as they sat passively, and he felt no resistance. They had the stiffness of a rag doll. Based on the path of the bullets, he realized that they all suffered damage to their cerebellum, the minibrain sitting below the back of the brain and attached to the brain stem. His localization was laser accurate. Even today, when a patient arrives at a neurology clinic with this constellation of clumsy movements, it is almost certain that she has something wrong with her cerebellum. Clearly, this structure plays a central role in shaping movement. It adds stiffness, but only as much as is needed to make a movement stable. It minimizes error, so actions are precise. It gets rid of extraneous muscle contractions, and it improves efficiency.

Since Holmes wrote his paper, the neuroanatomic connections of the cerebellum have been closely studied, providing a glimpse into how the organ might actually work. There is a massive trunk line of inputs from the brain to the cerebellum providing continuous real-time information about what the brain is up to. For example, a motor command that directs the arm to scratch the nose will descend toward the spinal cord and thence to the muscles. It also sends a copy into the cerebellum. In parallel, the cerebellum also receives all sorts of information about the position of the nose and the finger doing the scratching, tension across the arm joints, and the cognitive state of the scratcher. Integrating this information, it projects signals back to the motor cortical areas that modulate the ongoing scratching action. In essence, it works as a comparator, analyzing what the motor commands are initiating relative to what the brain actually wants to be doing.[14] As such, the cerebellum sits in a powerful position to adjust any and all movements from moment to moment, minute to minute, even year to year. It is another kind of critic of the motor command coming from the acting brain. Other critics in the brain, such as the basal ganglia, provide intermittent feedback, like a coach giving advice at the end of the game. In contrast, the cerebellar critic likely provides continuous feedback. It is like a coach shouting nonstop comments at a beleaguered athlete.

You can readily test your cerebellum. Put a heavy coat on and the sleeves will weigh down your arms. It should make your arm sag when you try to itch your nose. But there is no sag. The cerebellum detects the change and adjusts the gain of the motor command, exhibiting extremely swift adaption. If you had a cerebellar injury (which you can simulate by drinking far too much alcohol, a notorious inhibitor of the cerebellum), your arm would sag and would also wildly overcorrect in a flurry of jerky movements.

The cerebellum learns and remembers things about action that are fundamentally different from what the rest of the brain remembers. Synergies, skills, motor programs, and tool use are all functions stored in the cortex of the brain. In contrast, the cerebellum keeps track of a massive list of comparatively minor adjustments or tweaks to each movement to make them work well under a variety of conditions. Just think of a gymnastics coach fussing over where an athlete should point his toes while performing a triple flip. The cerebellum is making these adjustments throughout life, and its integrity is a good predictor of how well a person preserves finesse of movement as she ages.[15]

This distinction between what the cerebellum stores and what the rest of the brain remembers about movement became overwhelmingly clear to me when I was a young research fellow. My officemate was a young cardiologist who traveled home for a holiday. He developed high blood pressure and an artery exploded in his cerebellum. The ensuing blood clot was so large that it put his life in jeopardy. Neurosurgeons urgently opened up the base of his skull and evacuated the clot and the fragments that had once been his cerebellum. It took him two months to recover and return to the United States, only this time without a cerebellum. He moved much the way Holmes had described. A wide gait. Jumpy eye movements. Irregular control of fine movements. But he still knew how to do everything: cook, dress, bathe, type, write, play music, kick a football. He was just clumsy, challenged to move with grace and efficiency without a cerebellum to calibrate his every movement. He is now chairman of a cardiology department at a large university medical center.

If the brain or cerebellum is directly shaping movement based on how efficient the action is, then it ought to know how much fuel is being burned at any given moment. One way it might secure these measurements is from sensors located in the arteries of the neck that normally detect pressure, acidity, oxygen, and carbon-dioxide levels.[16] This information is relayed to the brain stem, where a host of reflexes adjust blood flow to the brain, blood pressure, heart rate, and respiration.[17] But are these measurements also used to keep track of fuel consumption? It is certainly plausible. These brain-stem nuclei have a direct connection to the cerebellum.[18] The cerebellum could constantly shape movement by tracking the energy costs from the information provided by the neck sensors. While plausible, this supposition is likely wrong. Max Donelan and his students tested whether changes in the amount of oxygen or carbon dioxide in the blood would influence how quickly people walked.[19] Participants wore a mask through which the experimenters controlled the concentration of inspired gas.

They decreased the levels of oxygen and increased carbon dioxide: exactly what happens when the body is working harder. If the cerebellum were (falsely) detecting an increase in fuel being burned, it should adapt movements to try to find a more efficient walking pattern. The participants initially reacted to the gas changes by slowing their cadence and breathing rate; however, they soon returned to their normal manner of walking. Even after extensive exposure to the simulated higher cost of transport, they showed no sign of changing their walking. Thus the brain's measurement of blood gases, one of its best ways to track fuel consumption, is probably not used for shaping movement at all. And so, if it is optimizing movement efficiency, it is doing it in an indirect, ingenious way that has yet to be discovered.

There are many other kinds of costs that the cerebellum might track instead of or in addition to efficiency. For many other tasks, like eye movements, pointing, throwing, jumping, or manipulating objects, inaccuracy can be extremely costly. It makes sense in these cases that spatial precision rather than efficiency should be maximized. Indeed, there has been a long-standing examination of how the cerebellum might minimize movement error.[20] Decades' worth of laboratory experiments focusing on many different kinds of actions identified spatial error as the one and only kind of cost that the cerebellum might work with. Another source of information the cerebellum might use to optimize movements is smoothness.[21] The cerebellum could learn to shape actions by sensing abrupt changes of force within the joints, of motor commands to the muscles, or of body position and shape.

The problem with all of these calculations is that for most actions, more than one cost is being used to shape movement. When a person is walking, the cerebellum needs to shape movement to both maintain stability and maximize efficiency.

Whenever there is more than one cost function to juggle, there will be inherent trade-offs. Consider two drivers, one who is obsessed with minimizing the cost of transportation and another with minimizing his commute time. The former gently controls changes of speed, extending acceleration over time, smoothly depressing and releasing the pedal to maximize fuel efficiency. The latter driver likes to use what engineers refer to as "bang-bang control," stomping on the accelerator pedal, peeling out as fast as possible, and braking suddenly and hard, which can wreak havoc on the car and its passengers. The problem with relying only on smoothness to control arm movements or walking is the same as being stuck behind a hyperefficient, slow driver. The problem with maximizing arm or walking speed is the same as being stuck with an erratic, inefficient driver. Brains have an additional problem with bang-bang control. The stronger the motor command, the noisier the signal, undermining the ability to hit the right speed.[22] Most of us drive between these two extremes. Similarly, the cerebellum finds a happy medium between different costs, finding reasonable trade-offs. Each time the body is perturbed, whether by a heavy coat or walking over rough terrain, it figures out the appropriate compromise.

Heading for the second of three passes, I traversed a rocky plain covered in trees that had been knocked down by a massive snow avalanche during a previous winter. I tried to maintain a brisk pace but was forced to weave, dodge, and step over a jumble of snags and logs. Without two good eyes, I lacked stereovision and tripped a few times from misjudging the height of branches scattered across the ground. Over time, my movements became more graceful; I was moving on this terrain with some semblance of efficiency.

To test the impact of a rough terrain on energy consumption in the lab, some creative scientists built a giant treadmill with obstacles glued onto the belt.[23] The walker has to step agilely onto, over, or off of each fake rock as the treadmill rolls along. Not surprisingly, when people do this dynamic kind of walking, they burn more calories. What is really amazing is how little the net impact is: they use only about 5 percent more energy. The system is "robust": the cerebellum can learn near-optimal behavior in a variety of less-than-ideal conditions without excessive costs.* Backpacks test robustness in a different way, by taxing stability and efficiency. As the load increases, a person makes incremental adjustments to his posture so that his legs will continue to swing like pendulums. The more they swing, the less energy is consumed. However, once the pack weighs a third of a person's body weight, his legs spread out to maintain balance, and his body calls on auxiliary muscles that are not normally needed to sustain balance.[24] No problem for the cerebellum. It will continue to smoothly adjust the tension in these newly recruited muscles, maintaining an impressive level of efficiency.

Getting up the second pass took me too much time. With my monocular vision, I was inevitably going more slowly and less efficiently, probably a smart compensation by the cerebellum to keep me from tripping.[25] Making things worse, I picked the wrong chute to ascend, and after thirty minutes of climbing I had to give up when a wall blocked further progress. I backtracked to the bottom of the cliffs and tried a second chute that was inviting at the bottom but cruel near the top. It was loaded

*Robust optimal control is a way of formally characterizing a system that will maintain control at some reasonable level across a broad range of operation conditions. Think of a jet airplane, which needs to keep flying in winds, rain, and snow, whether empty or full.

up with sand and gravel. For every two steps up, I slid back down one step, slowly wearing myself out. After finally reaching the pass, I stumbled down a series of terraces to a barren lake, where I took a break by lying in some long grass. The sun and horizon were starting to head for each other. It was only midafternoon, but this was fall and I still had a long way to go. All the efficiency, grace, and smoothness in the world would be pointless if I found myself still hiking at dark. My headlamp battery was dead and it would be a moonless night. I needed to step up my pace a couple of notches. I needed to add some oomph.

Until that moment, I had been passing through the mountains using a default setting for my walking speed: steady and relaxed. Adding some oomph to my pace would require more than just smoothness and grace; the entire tempo and intensity needed to be amplified. To do that, I employed a circuit in the brain that is routed through the basal ganglia, a collection of nuclei entwined in a complicated circuit deep inside the brain. They receive input from the cortex and then, through the use of the neurotransmitter dopamine, they ultimately modulate activity back in the originating cortex. We know this circuit is essential for relating rewards for particular actions, facilitating good old reinforcement learning. What is more, there is overwhelming evidence that these nuclei play a role in regulating the vim and vigor of movement itself—in adding oomph, as it were.[26] The earliest notion that this circuit was important for regulating the intensity of movement emerged with efforts to understand the pathophysiological basis of Parkinson's disease. Patients with Parkinson's are ubiquitously slow, and the scaling of their movement fails to match the task at hand. Their handwriting remains clearly identifiable as theirs, but with progressing disease the letters become smaller and smaller, until a reader might require a magnifying glass. When they reach for an object out in front, their effort presumes a closer object. Their walking stride shrinks, eventually becoming a shuffle. The spontaneous movements of their face become subtle, like those of an emotionless poker player. There is nothing wrong with the motor commands for these movements, which can sometimes be tricked into normalcy by replacing missing dopamine in the basal ganglia, thereby restoring amplitude of action.

A parallel story supporting the scaling hypothesis was developed in neurophysiology laboratories. As a result of the recording of neurons in the basal ganglia of monkeys, different aspects of arm movements could be linked with the activity of individual cells. In transformative research, Mahlon DeLong showed that the basal ganglia cells typically fired late, after a movement had already started, and their intensity of firing correlated with the amplitude of movement and, to a lesser degree, direction of arm movement. The firing was unrelated to the initiation of movement or the selection of one movement over another.[27] Soon after, DeLong became chair of the Department of Neurology at Emory University and recruited me to develop methods to try to understand what the basal ganglia contribute to human motor behavior. I teamed with Rob Turner and Michel Desmurget, and we set out to determine if vigor could be linked to the basal ganglia. We went back and looked at the accuracy of reaching movements in Parkinson's disease patients more closely. When asked to point to a target under dead reckoning (they could not see their own hand, only the target) they showed perfectly normal accuracy in the direction of their reach. However, they were highly variable in how far they would reach. Not only did they underreach because of a lack of vigor, there was also enormous inconsistency in amplitude from trial to trial.[28] Another important observation from Parkinson's disease patients was that they could be tricked into generating correctly sized movements if the right conditions were set up. If asked to make a small movement, they would make an abnormally tiny one. If asked to make a large movement, they would make a perfect medium-sized one. In other words, the problem was one of matching the amount of vigor with the unique requirements of the action.

But Parkinson's disease is also a complicated neurodegenerative disease with changes occurring throughout a patient's subcortical brain. To really make the case that the basal ganglia regulate movement amplitude it was essential to look for additional evidence. We also tested if the same sort of scaling could be taking place in healthy people. We used a PET scanner and injections of radioactive water to measure relative changes of blood flow in the brain as healthy participants performed a variety of simple motor tasks. In one trial, they were given advanced information on a particular direction to reach, but they did not know how far they were going to reach. In another, they knew how large of a reach they needed to make, but not the direction. When planning distance (but not direction), there was far more activity in the basal ganglia, consistent with its role in setting the vigor of the movement.[29] In two other experiments, we had subjects track a moving target with a giant joystick that required them to use the large muscles of the shoulder.[30] In one case, we varied how quickly they had to move (slow, medium, or fast), and in another, we varied how big the movements needed to be (small, medium, or large). In either case, there was more activity in the basal ganglia for fast or large movements. An even simpler concept of movement amplitude is isometric force. Here, David Vaillancourt showed in an extensive series of imaging experiments that the basal ganglia will vary their intensity in proportion to the amount of force a person uses to squeeze a rigid handle.[31] Putting all this evidence together, we became convinced that a major function of the basal ganglia is to set the proper level of vigor for a given movement in a particular situation. Collectively, we preferred the term "vigor" because it allows actions to be scaled in many ways, whether it is by changing the amplitude, velocity, scale, force, or acceleration of movement.[32]

Since our imaging studies of the basal ganglia, a theory describing the relationship between vigor and reward, the two known functions of the basal ganglia, has emerged. The theory, in its simplest form, boils down to relating how hard a person is willing to work to get a reward, with the basal ganglia being the key node for relating energy expenditure—*i.e.,* the oomph—and the importance of the reward. Supporting this theory, patients with Parkinson's disease have a hypersensitivity to cost of movement.[33] They will scale a movement only to the right size—that is, throw enough oomph into the task—if the rewards are big. Since most everyday actions afford little reward, for a patient with Parkinson's disease, the cost of a given movement usually doesn't seem to be worth the energy.[34] This complements the theory of what the cerebellum is doing in parallel: trying to maximize the efficiency of an action irrespective of the vigor or reward.

The trade-offs between effort and reward become obvious for anyone deciding what to bring on a hiking trip, whether to the wilderness or to a local park. When I prepare for a mountain adventure, I stuff my backpack with food and gear. Each pound of weight undermines a little bit of the stability of walking and increases the cost of transportation. Even so, a little more weight might allow for a delicious dinner or even a cocktail. Is the extra burden worth it? Vigor offers an alternative consideration: sometimes it is invaluable to be able to go farther faster and lighter, particularly to gain access to remoteness. Is the vigor worth it? Would I have a better time hiking like the tortoise or the hare? In the end, the answer in my case is simple. Bring better food and a cocktail and move like a tortoise. The reason has to do with the shape of the human body and the almost magical biomechanics of bipedal walking. On flat ground a person can readily carry up to a third of her body weight almost indefinitely with only a small increase in the cost of transport or perceived effort. Still, walking is only efficient when slow, and that efficiency quickly declines as a person speeds up.[35] And walking fast feels like work. Even the best long-distance hikers will admit that their success is not because they walk much faster than normal, it is because they walk for more hours each day.

That was how I had planned my trip a week earlier, but now I was watching the sun's transit with a different appreciation of value. Slow and heavy was out. I would be rewarded mightily if I got out of the mountains before dark or punished cruelly if I waited until first light, my failing vision aggravating my situation. I chose to expend as much energy as I could to go as fast possible, whatever the cost. I downed a late second lunch and started circling the lake. Soon I was lucky enough to pick up a small trail. I commenced a steady jog, using hiking poles to propel me uphill and to add lateral stability. The calculus underlying all my trade-offs of speed, stability, efficiency, and grace was performed without my giving it a single thought. All I had to do was sustain the tempo.

10

Of One Mind

Mind is everything. Muscles are pieces of rubber.
All that I am, I am because of my mind.

—PAAVO NURMI[1]

GETTING OUT BEFORE dark was starting to feel like a pipe dream as I realized how far I still had to go. It's one thing to tell yourself to go quickly, it's quite another to sustain that effort over an entire day. The sun on my back felt oppressive. A long, thin shadow threw itself out in front, a constant reminder of how late it was getting. I chased my shadow as fast as I could, but fatigue began to creep into my movements, stiffening me—melting away any semblance of grace. The fatigue was insidious, subverting my belief that I could make it out in time.

Fatigue. It creeps in from the darkest recesses of the mind, a fatalistic emotion that turns a person into a bleak pessimist. I could feel it, yet I could also fight back. I could still will myself forward, struggling to make it over the final pass. I picked up a large hiking trail with easier footing; I could take long, fast strides. I felt like I was almost flying. However, inside my head there were two angels battling it out. One made me feel like I would give up at any moment. It made me take notice of every flat-topped boulder I passed, each beckoning me to sit down and take a break. My better angel shouted a pep talk of perseverance, like an overenthusiastic motivational speaker.

The search for the source of fatigue reflects an old and familiar dichotomy: fatigue is either a product of the mind or a signal coming from the body. The issue impacts athletic training, exercise methods, and approaches to wellness adopted by medical science. The recognition that there could be at least two sources of fatigue—one mental, the other muscular—can be traced to the wonderfully engaging investigations of Angelo Mosso, a professor of physiology at the University of Turin. His fast-paced 1891 book *La Fatica* (*Fatigue*) begins, of all things, with a chapter on hummingbirds and their allegro wingbeat. He quickly shifts to the study of migratory birds and carrier pigeons and their capacity to travel thousands of miles. He poses the obvious question: Why don't these other species ever get fatigued?[2] He has a hunch there is something special about their muscles. He struggles to understand how muscles actually work but gets sidetracked by topics such as rigor mortis. He was groping for physiologic explanations of what makes muscles go and what can interfere with their power. His quest leads him to experiments where ground-up muscles from fatigued frogs are injected into dogs.* He analyzes exhausted children working in sulfur mines and then victims of writer's cramp. His search culminates in self-experimentation with an invention that he names the ergograph. The gadget would hold his forearm in place with the palm up, fingers outstretched. He would attach a weighted string to his middle finger and do curls with the finger: repeated flexions and extensions until the finger fatigued and the contractions petered out. What made the muscles stop functioning? A loss of nerve, something in

* Apparently, the dogs became tired. But who knows what the intravenous frog injection was really doing to them.

the muscle, or perhaps some sort of compound released by the muscle? He tries the ergograph on colleagues and discovers dramatic individual differences: Professor Aducco can sustain contractions for longer than Dr. Maggiora. Each of them practices frequently, and after seven years, Professor Aducco can sustain many more contractions: his muscles are gaining endurance. Mosso shifts to a focus on possible substances that might cause the fatigue. He notes that a frog muscle that has been electrically stimulated to the point of fatigue can be reconstituted by rinsing the muscle in salt water, suggesting either that it needs salt or that he is washing away a toxin. Finally, his greatest insight comes as a spectator at the rowing races on Lake Como. He monitors the breathing of the rowers and notes that respiration of the oarsmen increases from 14 to an almost breathless 120 per minute. He puts two and two together: what is impairing the muscles has something to do with a lack of oxygen. He brilliantly forecasts what would become, twenty years later, the canonical muscle-based theory of fatigue.

The muscle theory was eventually codified by Archibald Vivian Hill at University College London in 1923.[3] Hill too recognized the importance of oxygen and noted that breathing supplemental oxygen could enhance motor performance. Hill was also a self-experimenter; he went out to a short track and ran laps at different speeds corresponding to a pace of a 10-, 5-, and 3.75-minute mile. He measured the amount of lactate produced at each pace and noted how it skyrocketed with the fastest pace, when he was out of breath. Hill's explanation of fatigue: oxygen levels run low at peak performance, lactic acid is produced by anaerobic metabolism, and lactate acts as a poison that inhibits muscle contractions. Read any sports exercise textbook from the 1930s to today and this account dominates; circulating poisons shut muscles down.[4] Some of Hill's contemporaries made the case that the brain was probably sensing one or more chemical signals coming from the muscles and these triggered the feeling of tiredness. But Hill was less interested in the mentality of fatigue than he was in searching for a biochemical explanation. His account is a classic form of feedback control, similar to the way that many hormones in the body are regulated. Much like a wall thermostat, they respond to something sensed and respond accordingly. More than anyone, Hill established this reactive view of fatigue. With that in mind, the best training strategy for an athlete would be to build muscles that work longer and harder before the toxins are released. The problem with Hill's mechanism for fatigue is that it doesn't always make sense. Even at peak exertion, when an athlete is engaged in maximal effort, recent studies show that only about two-thirds of available muscle fibers are being used.[5] A lot of power seems to be kept in reserve. As I pushed onward through the mountains, I was going fast but nowhere near peak intensity. I had all the oxygen I needed and it is doubtful I had any measurable bump in lactate. While there might be molecules other than lactate circulating around and impacting my brain, it is still not clear what they might be. I was probably getting tired for some other reason.

Angelo Mosso had earlier proposed a different theory of fatigue. The brain might be fatigue's chief cause, irrespective of lactate or other molecules released by the muscles. In a chapter titled "The Law of Exhaustion," he presciently recognizes that fatigue could be an emotion engendered by the brain—one that serves a protective purpose: "The fatigue increasing more rapidly than the amount of work done saves us from the injury which lesser sensibility would involve for the organism."[6] Like Hill would later conceptualize, Mosso was proposing a type of reactive control that was based on feedback. Something in the brain was sensing the exertion and responding by limiting output, by generating fatigue. In engineering parlance, it acted as a governor to keep the driveshaft of an engine from spinning too fast. Mosso's proposal that there could be two sources of fatigue, generated by either brain or muscle, was acknowledged in the 1930s.[7] Mostly, though, the notion that fatigue could be a brain-derived emotion, one that could be further amplified by whatever was released by the muscles, was pretty much ignored for the rest of the century.

Remarkably, it really wasn't until this past decade that fatigue researchers seriously reconsidered how the brain could produce this intense emotion. And with that, the biggest breakthrough was seeing the proactive nature of fatigue, which anticipated rather than reacted to the state of the body. To see how easily fatigue can behave as an emotion, do some self-experimenting and manipulate your own mind to overcome it. For example, listening to music can greatly reduce your sense of fatigue.[8] Every exercise gym in the world is pumping in tunes to make you work longer and harder. One of the most insightful emotional manipulations of fatigue is to use a clock that runs too slowly.[9] The slow clock can trick you into dampening the development of fatigue. The clock effect is particularly important because it shows unequivocally that the brain is making a prediction about how long a physical activity should take for you to finish. In this light, fatigue is an emotion that is used to regulate both intensity and expenditure over some underlying period of time so that you can reach a particular goal with some reserve and without injury.[10] One can see this predictive calculus at work when an athlete is told to just run, without any idea of how long he will have to go. With this uncertainty, he will pick a conservative pace until he becomes aware of the finish, at which point he will accelerate.[11] Mosso was right. Fatigue is a strong emotion meant to keep a person from hurting himself. What Mosso failed to recognize was that the feeling is proactive: the brain anticipates the future and uses a powerful emotion to regulate intensity of effort.

The problem with a proactive emotion like fatigue is that, in general, the default setting is *extremely* cautious. It seems to be an emotion better suited for Ötzi, as if the brain is calibrated to keep a steady pace that will last for days, not hours. Knowing that fatigue can be, at least in part, a proactive "emotional" process has allowed elite athletes to engage in an internal battle, to push through their self-imposed mental barrier to improve performance. They recognize that winning the battle takes extensive practice and experience to learn the calculus of exhaustion. Physical intelligence learns how much fatigue to ignore.

I knew all that as I started to get tired on my trek. I knew that my brain was making a prediction about a finish line that was completely wrong. It was slowing me down to guarantee I could keep going all day long, for days on end, as I had been doing for the prior week. I had to recalibrate and set a faster pace. One that was going to hurt.

The other voice in the mind, the one resisting fatigue, is also powerful—an emotion based on a sense of urgency and perseverance. But where does this drive to persevere originate: the ability to grind, to push onward, to get through a tough situation? This impetus is not reserved for elite athletes. It is a feeling we can all draw on. So many difficult physical actions in real life need this stick-to-itiveness.* Only recently, a possible

*This capacity for working long and hard at a specific problem shouldn't be confused with the personality trait grit, which became a darling of the TED Talk circle but has largely been debunked as no more than a complicated recasting of other personality traits. See Crede, M., Tynan, M. C., Harms, P. D. "Much ado about grit: A meta-analytic synthesis of the grit literature." *J Pers Soc Psychol* 113 (2017): 492–511.

source of this tough-minded perseverance has come to light. At Stanford University, an epilepsy surgery team stumbled upon at least one place in the brain where this feeling might be generated. Like Wilder Penfield and countless neurosurgeons since, they stimulated the brain of their awake patients while searching for a source of epilepsy that could be removed.[12] In the course of stimulating the anterior cingulate cortex, two of their patients reported a remarkable series of feelings. The verbatim conversation tells the story: The examiner asks the first patient, "Any change in your emotion and mood?" The response is exceptional. "Not in my emotion, but my mood, I started getting this feeling like . . . I was driving into a storm. That's the kind of feeling I got. Like, almost like you're headed toward a storm that's on the other side, maybe a couple of miles away, and you've got to get across the hill and all of a sudden, you're sitting there going how am I going to get over that, through that? And that's the way my brain started functioning." After another stimulation, the patient notes his chest feels like it is pounding, but makes an important distinction, "like you're a football player getting ready to go out and make his first touchdown for the season or something; it's not that type of thing. It's more like this thing of trying to figure out your way out of, how you're going to get through something. It's not a matter of how you're going to production-wise do something . . . Let's say . . . if you knew you were driving your car and it was . . . one of the tires was half flat and you're only halfway there and you have no other way to turn around and go back, you have to keep going forward . . . That type of a, you know, feeling you have. You're like, you're like (pats chest) am I gonna, am I gonna to [*sic*] get through this? Am I gonna get through this?" And asked if it was a negative or positive feeling he reported, "It was more of a positive thing like . . . push harder, push harder, push harder."

Here was a beautiful description of the mental state a person could tap into to overcome the limits imposed by fatigue.

The modern athlete's ability to persevere against the emotion of fatigue has resulted in a new and widespread phenomenon at the end of long-distance races: the finish-line collapse. This is not a slow and graceful fall to the ground. No, the modern athlete finishes with an instantaneous flop to the track like a Raggedy Ann doll that someone flung across the finish line. This dramatic ending has become routine in cross-country skiing, triathlons, and long-distance running. Once down, the athlete lies on the ground puffing desperately to catch his breath, like Mosso's rowers. Within a minute or two he is back up on his feet, hugging and laughing. Clearly, the fact that the collapse happens after he crosses the finish line shows that his training has calibrated the battle between the two emotions to perfection. As he crosses the finish line, perseverance concedes to fatigue. What is hard to recognize is the viciousness of the internal struggle, the intensity of discomfort that fatigue can generate, and the years of training that are needed to wage battle against it successfully.

The region of the anterior cingulate cortex that was associated with perseverance in the epilepsy patients is an anatomic sweet spot. For one, it is a major hub that interacts with the premotor areas that organize physical action.[13] Given the location, it should be no surprise that perseverance and physical drive are closely related. Furthermore, the anterior cingulate cortex is well recognized as embedded within a larger network of frontal lobe regions that are involved in many forms of cognition, particularly decision-making tasks comparing reward and punishment.[14]

When I started walking on the day in question, the reward for me was getting out before dark; the punishment was not making it out and spending a night in a dry camp without food or water. Should I accept or reject this trade-off? In recent experiments in our laboratory, we have been measuring brain activity in regions including the anterior cingulate cortex as people accept or reject analogous conflicting offers. Consider one experiment: a person is offered a certain amount of money at the cost of having to ride a bicycle at a particular intensity for one minute. A typical participant will accept high monetary offers for low-intensity workouts and reject low monetary offers for high-intensity bicycling. Between those extremes, each participant in the experiment has a distinct liminal zone for switching between accepting and rejecting and vice versa.[15] The cingulate cortex and nearby frontal cortex track an individual's subjective valuation of each offer. It does the calculus for the take-it-or-leave-it decision. How might fatigue influence this process?

With respect to my own situation outdoors, one might think that fatigue would be the punishment. But that is a mistake. When I was deciding to start my fast walking that morning, I was comparing the benefits of getting out before dark or not with the cost of pushing myself to a certain physical intensity. I wasn't fatigued at all. If I had been fatigued that morning, I would have been far more pessimistic in believing I could actually go fast enough to get out in time. In other words, fatigue is acting on the decision-making apparatus, shifting the boundary between what to accept or what to reject as an action. One highly plausible way that fatigue could do this was discovered by my colleague Ann Graybiel and her team at MIT. They have achieved a deep neurophysiologic dive into the anterior cingulate cortex of monkeys by recording neurons as the animals make analogous take-it-or-leave-it decisions about reward and aversion.[16] Even at this microscopic resolution, one can observe how individual neurons will track the relative value of different offers. What is even more remarkable is that when the researchers electrically stimulated this brain area, they could turn an animal into a pessimist.[17] An offer that was previously acceptable was no longer good enough. Substitute fatigue for the stimulation and you will likely get the same effect. In related research in patients with severe depression, it is clear that this region of the cortex is also intertwined with circuits that powerfully manipulate a person's mood, and it is thought to play a critical role in generating depressive symptoms.[18] Pessimism can run rampant in the depressed mind, profoundly altering life choices. Thus we are beginning to see how all these seemingly distinct cognitive processes are intimately interrelated, binding choice, mood, perseverance, and fatigue.

I crossed the final pass. It was only at about 11,800 feet above sea level, but I had been mushing as fast as I could to get there. As I slowed to pull out a water bottle, I staggered a bit and felt a moment of dread, fearful of the slow collapse that can bedevil athletes before they reach the finish line. The slow collapse occurs when people push themselves to near catastrophic system failure. Remarkably, this can even happen without their feeling fatigue. This condition is provoked by lack of preparation for the length of the journey, overheating, concomitant sickness, or choice of an unsustainable pace. At some point, they run out of energy or sodium, they tax their muscles to exhaustion (as Archibald Hill would have predicted), rendering themselves susceptible to heatstroke or, even worse, cerebral or cardiac thrombosis. A tip-off that things are starting to go sour for an athlete is an unstable gait. That was my warning sign as I staggered around at the final pass.

In the face of impending catastrophic system failure, the brain of an athlete who continues to push onward initiates a remarkably predictable set of proactive maneuvers to protect itself as the calamity unfolds, codified as the "Foster collapse positions."[19] The athlete assumes the "Early Foster" collapse position with his hips slightly flexed and his head down. As disaster mounts, he starts to shuffle and assumes a "Half Foster," with the trunk and head bent and parallel to the ground. With the head held low, postural support and the coordination of propulsion are severely compromised. It is hard to watch an athlete toiling forward in this position, but it is a beautiful example of proactive control, the brain quietly reshaping posture and motion to keep the head down so it can sustain cerebral circulation. One doesn't need a blood-pressure cuff to see what is going on here. Remarkably, the athlete seems to be oblivious to these warning signs of imminent system failure. He pushes on! Soon, he will fall to the ground and crawl forward on knees and elbows, assuming the "Full Foster" with his head down and lower than the hips, further maintaining perfusion of the brain. Upon final collapse he remains prone for an extended period. There is no energy left to roll over into a more comfortable supine position. When Saint-Exupéry collapsed at the feet of the nomads in the Libyan desert, he remained prone like this as they lifted his head and slid a pan of water under his face for him to lap. Like so many athletes who have become prone since, he chose to continue on until the very end. Needless to say, I didn't want to end up with my face in a pan of water.

As I started to stagger, I was no longer managing a tension between perseverance and fatigue; I was in an outright battle between perseverance and stress. The slowly collapsing body is pounded with physiologic stress. Cellular metabolism, neuronal transmission, and cardiac pumping are all approaching failure. The striking feature of the slow collapse is the ease with which I could override obvious signals that something was very wrong with my body. This is a familiar dualism: a fundamental instance of mind over matter. The body (and most of the brain, for all that) functions as a galley slave, desperately rowing a trireme onward. The mind is a slave driver, whipping the rower relentlessly, uninterested in their fate. It is up to the slave to fend for himself, and a person can literally row himself to death. The stresses are sensed and reacted to by the body, while the mind remains oblivious. As the stress mounts, the deeper structures of the brain like the hypothalamus (but not the mind) react by flooding the circulation with stress hormones to maintain, more than anything else, blood pressure and cellular metabolism. Other brain-stem nuclei drive the heart harder and adjust the blood vessels to keep the blood pressure up. Familiar stress hormones like cortisol work more slowly and broadly, influencing metabolism throughout the body and enhancing the eventual recovery of the body from a major stressor. All of these well-established physiologic responses to stress are based in part on feedback control: the brain senses trouble and responds to it.

This traditional physiological model of stress is reactive. A person sees a bear, adrenaline floods the person's circulation, stimulating the heart so she can run or fight. However, it is increasingly recognized that the mind also shapes stress responses. The mental brain plays a nuanced and proactive role in shaping the body's visceral responses, particularly those occurring on fast timescales. These responses are mediated in large part by the sympathetic nerves of the autonomic nervous system and by hormones such as adrenaline that drive the heart and adjust blood pressure by anticipating what the body will need in the near future. For example, if you don't want to pass out each time you stand up, it helps to have a mechanism that starts to adjust your blood pressure before you stand up.

One of the earliest insights that the mind is important for mediating the organs swiftly, particularly changes involving the heart, was proposed by the physiologist Walter Cannon. In a classic paper titled "'Voodoo' Death," published in an anthropology journal, he reviewed the countless examples of people from many different cultures being literally scared to death by voodoo, witchcraft, or curses.[20] He argued that voodoo death was real and caused by a shocking emotional stress that was sufficiently strong to drive the sympathetic nervous system to a point where the heart was damaged or blood pressure collapsed. He was fuzzy about how the emotions and the brain might be directing the sympathetic nervous system to do this; the anatomy was still unknown. Nevertheless, he clearly recognized a role for the mind in shaping the behavior of the heart and blood vessels. While voodoo death has become rare in modern cultures, sudden cardiac death after brain injury is now well recognized among neurologists.[21] Patients with a stroke that damages the cortex of the brain are at greater risk for dying suddenly from heart failure. The implication of this is that the cortex itself can be directly engaged in quick stress responses, particularly those involving the heart and blood vessels. But where in the cortex is this signal coming from?

There are two lines of evidence that will eventually need to be reconciled. The standard view, based on extensive studies of patients who have suffered strokes and brain imaging of healthy people, is that the regulation of the heart is mediated through the insula: a hidden island of cortex located underneath the temporal lobe. In healthy subjects, changes in heart rate can be correlated with insula brain activity measured by functional MRI scans. And in patients with stroke, damage here can lead to catastrophic heart failure, presumably because of brief overstimulation of the heart.[22] Recent anatomic studies by my colleague Peter Strick suggest that the insula serves as a control center to the parasympathetic nerves. These are the nerves that actually brake the heart and allow us to rest and digest. There is also good evidence that this area senses feelings from the viscera of the body. It is monitoring heart rate, respiration, and activity of the gut, making sense of what is going on.[23] Damaging this sensing might also eliminate feedback control and lead to runaway stimulation of the heart.

The alternative model, which we know less about, maintains that the cingulate cortex and other areas of the frontal lobe are key for connecting the mind to the body by proactively controlling fast-acting stress responses via the sympathetic nerves.[24] This is the system that drives the heart harder and enhances blood flow to the muscles. The evidence supporting this conclusion is quickly mounting. In recent studies of patients with epilepsy, when electrodes are placed in the deep portion of the anterior cingulate and a jamming signal is introduced, there is an obvious and dramatic decline in systolic blood pressure that is sustained until the stimulator is turned off.[25] Such fast changes could be sparked only by nerves, not circulating hormones. No other area of the human brain has ever been shown to manipulate blood pressure this reliably. What about the release of short-acting hormones like adrenaline from the adrenal gland? Textbooks classically show that the adrenal medulla and the heart are innervated by sympathetic nerves that can be traced back through a few relay stations to nuclei in the brain stem. But what part of the brain tells the brain stem when the adrenal should release hormones like adrenaline and drive the heart harder? Recent anatomic tract tracing studies show that the anterior cingulate is one of only a few areas that reach the adrenal medulla via the brain-stem relay station.[26] We suspect that this area will also be one of the dominant cortical sources that shape the sympathetic drive to the heart. In my laboratory we have been measuring the beat-to-beat intensity of the contractions of the heart, another way to look at sympathetic drive, while people are in an fMRI scanner.[27] We found that the anterior cingulate is one of the areas where the brain activity leads rather than lags the intensity of the cardiac drive: a clear case of proactive control.

These discoveries linking the mind, emotions, cognitive systems, and the core circuits for regulating basic body physiology completely change how one might think about the slow collapse. On first glance it seems to be a disconnection between mind and body, a failure of the person to recognize impending system failure. The new findings suggest there is no disconnection. The brain is not naïve when it comes to the body. The person stumbling onward with head down has made a hard choice to continue. The brain area that is involved in making this choice is also tracking the heart, tuning the heartbeats, adjusting blood pressure, and anticipating increasingly difficult physiologic demands as the various muscles and visceral organs are stressed to the point of failure. The Foster collapse positions are a beautiful example of the brain making predictions and adjusting posture to preserve circulation to the brain. The proactive role of the anterior cingulate in this visceral refinement is particularly salient. Recall that this is the same brain system that calculates the subjective value of action, perseverance, cardiac output, and blood pressure, as well as mood regulation. The interconnectedness of cortical systems for thinking, sustaining physical life, and setting tangible goals is hardwired into all of us. Insofar as physical intelligence is concerned, the distinctions between mind and body are illusory. Saint-Exupéry, crawling toward his own rescue in a Full Foster, had long since calculated the trade-offs between survival or not. His body knew exactly what it was doing and was more than willing to take the chance of a catastrophic, system-wide failure. I too knew full well the risk and reward of pushing onward, albeit with increasingly shaky legs.

—

Fortunately, I was able to keep my head up. There would be no Half Foster on this trip. As I trundled into the final switchbacks of the trail, zigzagging down a hillside covered in sagebrush, I started to feel as high as a kite. Even though I was nearing collapse, I had an intense sense of well-being, of the elegant interconnectedness of my actions, thoughts, and ability to incorporate all that I was experiencing. I had beaten myself up for a long week in a very rough world. Despite that, I had a sense that the week had already been worth it, that I was healthier and stronger, both cognitively and physically. I had taken a path toward well-being that is profoundly different from what I was accustomed to back home. There, I would have to stay fit with boring exercise. Fat would have to be shed through strict food choices. And if I am to believe the advertisements, the well-being of my mind would have to be preserved by exercising my cognitive skills with tedious computer games. None of these strategies make sense in the wilderness. For me, this divide between the physical and the cognitive, as well as the emphasis on training rather than engagement, describes a condition of modern society that saddens me, as it does not allow for the holistic recognition of the unity of a person's feelings, thinking, and physicality.

Throughout this book, I have made the case that the integrity of a person can be revealed through the intelligence of physical action. Intense physical experience, particularly in complex natural settings, places demands on the brain to learn and to be proactive, even as it refines action to allow for best performance. Over the years, I have been fascinated by the differences between neurology patients who have done well and those who haven't. Similarly, we all know people who age gracefully and others who don't. My neurology patients who do well derive their resilience from rich physical engagement. These patients don't just jog on a treadmill in the hope that their muscles will release life-enhancing molecules. They drive bulldozers, plant trees, learn to play the violin, make jewelry, remodel their house, or take dance lessons. They physically interact with their environment in ways far more intense than one might expect from a person with emerging disabilities. They are like Stephen Hawking, willing to travel to the South Pole just to see it. Indeed, there is growing evidence that engagement tasks are just as powerful, if not more powerful, than any other training strategy for improving the aging mind.[28] What particular engagement works best remains to be seen. However, from large-scale meta-analyses of aging it appears that physical engagement, particularly walking outdoors, is one of the top performers in sustaining cognitive health.[29]*

A rich and complex connection to the physical world demanding abilities that blur distinctions among mind, brain, and body; proactive decision-making regarding physiological events and action execution; physically challenging do-it-yourself projects in complicated situations that require ingenuity, strength, and a willingness to stray off the beaten path: these desiderata, in their variety and complexity of physical action, allow for the sustenance of enviable personhood. The challenge for our future, particularly as we more and more partake of a cocooned urban lifestyle, will be to find settings analogous to nature that will require sufficient complexity of physical intelligence and ensure the physical experiences we need to sustain our health and provide us with a sense of integrated well-being.

*The other top performers are good genes and having a rich social network.

Acknowledgments

My thanks to Phil Arnot, who, a very long time ago, ignited the flame that constantly draws me back to the wild; to David Lunn, who taught me when to laugh while getting buried in an avalanche; and to all the other great climbers who shared in foolhardy adventures. The book would never have happened if Mike Phelps hadn't shown me how to scan a brain, if John Mazziotta hadn't convinced me I could write, and if Rich Ivry hadn't pushed me to ask great questions. I am in debt to every patient who ever sought my help; they are the greatest teachers. The final push to do this project stems from Mike Gazzaniga's insistence that I should, and Max Brockman, for enthusiastically representing me. Becky Gazzaniga was invaluable for coaxing me into sending her early drafts. Finally, I am honored to have had the good fortune of Erroll McDonald at Pantheon as my editor.

Notes

Introduction

1. Quoted by Maggie Downs. "Rock around the clock." *Palm Springs Life Magazine,* December 31, 2015. In the late 1970s, Jim Bridwell transformed the world of rock climbing.
2. Quoted by Claudia Dreifus. "Life and the Cosmos, Word by Painstaking Word." *The New York Times,* May 9, 2011.

1. The Space We Create

1. From an interview in *USA Today,* January 7, 2013.
2. Holmes, N. P., Spence, C. "The body schema and the multisensory representation(s) of peripersonal space." *Cogn Process* 5 (2004): 94–105.
3. Graziano, M. S., Gross, C. G. "Spatial maps for the control of movement." *Curr Opin Neurobiol* 8 (1998): 195–201.
4. Brain, W. R. "Visual disorientation with special reference to lesions of the right cerebral hemisphere." *Brain* 64 (1941): 244–72; Oxbury, J. M., Campbell, D. C., Oxbury, S. M. "Unilateral spatial neglect and impairments of spatial analysis and visual perception." *Brain* 97 (1974): 551–64.
5. Gerstmann, J. "Some notes on the Gerstmann syndrome." *Neurology* 7 (1957): 866–69; Holmes, G. "Disturbances of Visual Orientation." *Br J Ophthalmol* 2 (1918): 449–68, 506–16.
6. Pandey, S., Tomar, L. R., Tater, P. "Pisa syndrome in a child with subacute sclerosing panencephalitis." *JAMA Neurol* 75 (2018): 255–56.
7. Karnath, H. O. "Pusher syndrome—A frequent but little-known disturbance of body orientation perception." *J Neurol* 254 (2007): 415–24.
8. Nardo, D., Console, P., Reverberi, C., Macaluso, E. "Competition between visual events modulates the influence of salience during free-viewing of naturalistic videos." *Front Hum Neurosci* 10 (2016): 320.
9. Fodor, J. A. *Modularity of Mind: An Essay on Faculty Psychology.*

Cambridge, MA: MIT Press, 1983; Gazzaniga, M. S. "Organization of the human brain." *Science* 245 (1989): 947–52.

10. Jackson, J. H. "Clinical and physiological researches on the nervous system. I. On the anatomical and physiological localization of movements in the brain." In *Selected Writings*. London: J. & A. Churchill, 1875.
11. Mesulam, M. M. "A cortical network for directed attention and unilateral neglect." *Ann Neurol* 10 (1981): 309–25.
12. Chen, T., Cai, W., Ryali, S., et al. "Distinct global brain dynamics and spatiotemporal organization of the salience network." *PLoS Biol* 14 (2016): e1002469.
13. Castiello, U., Umiltà, C. "Size of the attentional focus and efficiency of processing." *Acta Psychol* (Amst) 73 (1990): 195–209.
14. Eriksen, C. W., James, J. D. S. "Visual attention within and around the field of focal attention: A zoom lens model." *Percept Psychophys* 40 (1986): 225–40.
15. Turatto, M., Benso, F., Facoetti, A., et al. "Automatic and voluntary focusing of attention." *Percept Psychophys* 62 (2000): 935–52.
16. Jefferies, L. N., Gmeindl, L., Yantis, S. "Attending to illusory differences in object size." *Atten Percept Psychophys* 76 (2014): 1393–402.
17. Iwasaki, S. "Spatial attention and two modes of visual consciousness." *Cognition* 49 (1993): 211–33.
18. Beume, L. A., Kaller, C. P., Hoeren, M., et al. "Processing of bilateral versus unilateral conditions: Evidence for the functional contribution of the ventral attention network." *Cortex* 66 (2015): 91–102.
19. Nardo, Console, Reverberi, Macaluso. "Competition between visual events modulates the influence of salience during free-viewing of naturalistic videos."
20. Webb, T. W., Igelström, K. M., Schurger, A., Graziano, M. S. A. "Cortical networks involved in visual awareness independent of visual attention." *Proc Natl Acad Sci USA* 113 (2016): 13923–28.
21. Patel, G. H., Yang, D., Jamerson, E. C., et al. "Functional evolution of new and expanded attention networks in humans." *Proc Natl Acad Sci USA* 112 (2015): 9454–59.
22. Stawarczyk, D., Majerus, S., Catale, C., D'Argembeau, A. "Relationships between mind-wandering and attentional control abilities in young adults and adolescents." *Acta Psychol* (Amst) 148 (2014): 25–36.
23. Farrant, K., Uddin, L. Q. "Asymmetric development of dorsal and ventral attention networks in the human brain." *Dev Cogn Neurosci* 12 (2015): 165–74.
24. Sanefuji, M., Craig, M., Parlatini, V., et al. "Double-dissociation between the mechanism leading to impulsivity and inattention in Attention Deficit

Hyperactivity Disorder: A resting-state functional connectivity study." *Cortex* 86 (2017): 290–302.

25. Sripada, C. S., Kessler, D., Angstadt, M. "Lag in maturation of the brain's intrinsic functional architecture in attention-deficit/hyperactivity disorder." *Proc Natl Acad Sci USA* 111 (2014): 14259–64.
26. Asken, M., Christensen, L. W., Grossman, D. *Warrior Mindset: Mental Toughness Skills for a Nation's Peacekeepers.* Amazon Digital Services LLC, 2017.
27. Ranadive, V., Maney, K. The Two-Second Advantage: How We Succeed by Anticipating the Future—Just Enough. New York: Crown Business, 2011.
28. Russman, B. S., King, A. P., Sripada, R. K., et al. "Behavioral and neural correlates of disrupted orienting attention in posttraumatic stress disorder." *Cogn Affect Behav Neurosci* 17 (2017): 422–36.
29. Schäfer, J., Bernstein, A., Zvielli, A., et al. "Attentional bias dynamics predict posttraumatic stress symptoms: A prospective-longitudinal study among soldiers." *Depress Anxiety* 33 (2016): 630–39.
30. Badura-Brack, A. S., Naim, R., Ryan, T. J., et al. "Effect of attention training on attention bias variability and PTSD symptoms: Randomized controlled trials in Israeli and U.S. combat veterans." *Am J Psychiatry* 172 (2015): 1233–41.
31. Lutz, A., Slagter, H. A., Dunne, J. D., Davidson, R. J. "Attention regulation and monitoring in meditation." *Trends Cogn Sci* 12 (2008): 163–69.
32. Kaufmann, 2015.
33. Taren, A. A., Gianaros, P. J., Greco, C. M., et al. "Mindfulness meditation training and executive control network resting state functional connectivity: A randomized controlled trial." *Psychosom Med* 79 (2017): 674–83.

2. Surfaces

1. From the recording "Ask," by Morrissey and Johnny Marr, performed by the Smiths, Sire Records, 1986.
2. Uexküll, J. von, ed. *A Stroll Through the Worlds of Animals and Men: A Picture Book of Invisible Worlds.* New York: International Universities Press, Inc., 1957.
3. Gibson, J. J. *The Ecological Approach to Visual Perception.* Hillsdale, NJ: Lawrence Erlbaum Assoc., 1979.
4. Garm, A., O'Connor, M., Parkefelt, L., Nilsson, D. E. "Visually guided obstacle avoidance in the box jellyfish *Tripedalia cystophora* and *Chiropsella bronzie.*" *J Exp Biol* 210 (2007): 3616–23.

5. Foulsham, T., Walker, E., Kingstone, A. "The where, what and when of gaze allocation in the lab and the natural environment." *Vision Res* 51 (2011): 1920–31.
6. Mark, L. S. "Eyeheight-scaled information about affordances: A study of sitting and stair climbing." *J Exp Psychol Hum Percept Perform* 13 (1987): 361–70; Warren, W. H., Jr. "Perceiving affordances: Visual guidance of stair climbing." *J Exp Psychol Hum Percept Perform* 10 (1984): 683–703.
7. Fischer, M. H. "Estimating reachability: Whole-body engagement or postural stability?" *Hum Mov Sci* 19 (2000): 297–318; Rochat, P., Wraga, M. "An account of the systematic error in judging what is reachable." *J Exp Psychol Hum Percept Perform* 23 (1997): 199–212.
8. Fajen, B. R., Matthis, J. S. "Direct perception of action-scaled affordances: The shrinking gap problem." *J Exp Psychol Hum Percept Perform* 37 (2011): 1442–57.
9. Holmes, G. "Disturbances of Visual Orientation." *Br J Ophthalmol* 2 (1918): 449–68, 506–16.
10. Schaadt, A. K., Brandt, S. A., Kraft, A., Kerkhoff, G. "Holmes and Horrax (1919) revisited: Impaired binocular fusion as a cause of 'flat vision' after right parietal brain damage—A case study." *Neuropsychologia* 69 (2015): 31–38.
11. Lescorart, M. D., & Gallant, J. L. (2019). Human scene selective areas represent 3D configurations of surfaces. *Neuron* 101 (1), 178–92.
12. Adolph, K. E. "Specificity of learning: Why infants fall over a veritable cliff." *Psychol Sci* 11 (2000): 290–95.
13. Tayler, J. E., Witt, J. K., Sugovic, M. "When walls are no longer barriers: Perception of wall height in parkour." *Percept Psychophys* 40 (2011): 757–60.
14. Pezzulo, G., Barca, L., Bocconi, A. L., Borghi, A. M. "When affordances climb into your mind: Advantages of motor simulation in a memory task performed by novice and expert rock climbers." *Brain Cogn* 73 (2010): 68–73.
15. Handy, T., Tipper, C. M., Borg, J. S., et al. "Motor experience with graspable objects reduces their implicit analysis in visual- and motor-related cortex." *Brain Res* 1097 (2006): 156–66.
16. Mark, L. S., Vogele, D. "A biodynamic basis for perceived categories of action." *J Mot Behav* 19 (1987): 367–84.
17. Sperandio, I., Kaderali, S., Chouinard, P. A., et al. "Perceived size change induced by nonvisual signals in darkness: The relative contribution of vergence and proprioception." *J Neurosci* 33 (2013): 16915–23; Volcic, R., Fantoni, C., Caudek, C., et al. "Visuomotor adaptation changes

stereoscopic depth perception and tactile discrimination." *J Neurosci* 33 (2013): 17081–88.

18. Witt, J. K., Proffitt, D. R., Epstein, W. "Tool use affects perceived distance, but only when you intend to use it." *J Exp Psychol Hum Percept Perform* 31 (2005): 880–88.
19. Seok, S., Wang, A., Chuah, M. Y., et al. "Design principles for highly efficient quadrupeds and implementation on the MIT Cheetah robot." Presented at 2013 IEEE International Conference on Robotics and Automation (ICRA), Karlsruhe, Germany, 2013.
20. Raibert, M., Blankespoor, K., Nelson, G., Playter, R. "BigDog, the rough-terrain quadruped robot." *IFAC Proc Vol* 41 (2008): 10822–25.
21. Shkolnik, A., Levashov, M., Manchester, I. R., Tedrake, R. "Bounding on rough terrain with the LittleDog robot." *Int J Rob Res* 30 (2011): 192–215.
22. Seok, Wang, Chuah, et al. "Design principles for highly efficient quadrupeds and implementation on the MIT Cheetah robot."
23. Filitchkin, P., Byl, K. "Feature-based terrain classification for LittleDog." Presented at 2012 IEEE/RSJ International Conference on Intelligent Robots and Systems (IROS 2012), 2012.
24. Adolph, K. E., Cole, W. G., Komati, M., et al. "How do you learn to walk? Thousands of steps and dozens of falls per day." *Psychol Sci* 23 (2012): 1387–94.
25. Gelbard, R., Inaba, K., Okoye, O. T., et al. "Falls in the elderly: A modern look at an old problem." *Am J Surg* 208 (2014): 249–53.
26. Giladi, A. M., Shauver, M. J., Ho, A., et al. "Variation in the incidence of distal radius fractures in the U.S. elderly as related to slippery weather conditions." *Plast Reconstr Surg* 133 (2014): 321–32.
27. Der Ananian, C. A., Mitros, M., Buman, M. P. "Efficacy of a student-led, community-based, multifactorial fall prevention program: Stay in Balance." *Front Public Health* 5 (2017): 1–12; Sherrington, C., Michaleff, Z. A., Fairhall, N., et al. "Exercise to prevent falls in older adults: An updated systematic review and meta-analysis." *Br J Sports Med* 51 (2016): 1750–58.
28. Konczak, J., Meeuwsen, H. J., Cress, M. E. "Changing affordances in stair climbing: The perception of maximum climbability in young and older adults." *J Exp Psychol Hum Percept Perform* 18 (1992): 691–97.
29. Kuehn, E., Perez-Lopez, M. B., Diersch, N., et al. "Embodiment in the aging mind." *Neurosci Biobehav Rev* 86 (2018): 207–25.
30. Gelbard, Inaba, Okoye, et al. "Falls in the elderly: a modern look at an old problem."
31. Nicklett, E. J., Lohman, M. C., Smith, M. L. "Neighborhood environ-

ment and falls among community-dwelling older adults." *Int J Environ Res Public Health* 14 (2017): 175.
32. Roubik, D., Cook, A. D., Ward, J. G., et al. "Then we all fall down: Fall mortality by trauma center level." *J Surg Res* 217 (2017): 36–44.
33. Chippendale, T., Raveis, V. "Knowledge, behavioral practices, and experiences of outdoor fallers: Implications for prevention programs." *Arch Gerontol Geriatr* 72 (2017): 19–24.
34. Giladi, Shauver, Ho, et al. "Variation in the incidence of distal radius fractures in the U.S. elderly as related to slippery weather conditions."

3. Shaping the Self

1. Ovid. *Metamorphoses*. Trans. David Raeburn. London: Penguin Books, 2004.
2. Todd, J. "The syndrome of Alice in Wonderland." *Can Med Assoc J* 73 (1955): 701–4.
3. Bell, C. *The Hand; Its Mechanism and Vital Endowments as Evincing Design*. London: William Pickering, 1834.
4. Duchenne, G. "*De l'ataxie locomotrice progressive.*" *Arch Gén de Méd* 5 (1858): 641–52.
5. de la Tourette, G. *Études cliniques et physiologiques sur la marche: La marche dans les maladies du système nerveux; Étudiée par la méthode des empreintes*. Paris: Impr. Noizette, 1886.
6. Goetz, C. G., Bonduelle, M., Gelfand, T. *Charcot: Constructing Neurology*. New York: Oxford University Press, 1995.
7. Proske, U., Gandevia, S. C. "The proprioceptive senses: Their roles in signaling body shape, body position and movement, and muscle force." *Physiol Rev* 92 (2012): 1651–97.
8. Head, H., Holmes, G. "Sensory disturbances from cerebral lesions." *Brain* 34 (1911): 102–254.
9. Goldmann, J.-P., Braunstein, B., Sanno, M., et al. "Fast-roping: Potential consequences of vibrations for sensation and regulation of movement." *J Spec Oper Med* 14 (2014): 80–83.
10. Lackner, J. R. "Some proprioceptive influences on the perceptual representation of body shape and orientation." *Brain* 111, pt. 2 (1988): 281–97.
11. Botvinick, M., Cohen, J. "Rubber hands 'feel' touch that eyes see." *Nature* 391 (1998): 756.
12. Lopez, C., Lenggenhager, B., Blanke, O. "How vestibular stimulation interacts with illusory hand ownership." *Conscious Cogn* 19 (2010): 33–47.
13. Dave Wolf describes this experience in detail in a Radiolab episode titled "Darkness."
14. Desmurget, M., Rossetti, Y., Jordan, M., et al. "Viewing the hand prior

to movement improves accuracy of pointing performed toward the unseen contralateral hand." *Exp Brain Res* 115 (1997): 180–86.

15. Joseph, R. "Fetal brain behavior and cognitive development." *Dev Rev* 20 (2000): 81–98.
16. von Hofsten, C. "An action perspective on motor development." *Trends Cogn Sci* 8 (2004): 266–72.
17. Rochat, P. "Self-perception and action in infancy." *Exp Brain Res* 123 (1998): 102–9.
18. van der Meer, A. L. "Keeping the arm in the limelight: Advanced visual control of arm movements in neonates." *Eur J Paediatr Neurol* 1 (1997a): 103–8.
19. Mathew, A., Cook, M. "The control of reaching movements by young infants." *Child Dev* 61 (1990): 1238–57.
20. Lockman, J. J., Ashmead, D. H., Bushnell, E. W. "The development of anticipatory hand orientation during infancy." *J Exp Child Psychol* 37 (1980): 176–86.
21. Cole, J., Waterman, I. *Pride and a Daily Marathon*. Cambridge, MA: MIT Press, 1995.
22. Diedrichsen, J., Wiestler, T., Krakauer, J. W. "Two distinct ipsilateral cortical representations for individuated finger movements." *Cereb Cortex* 23 (2013): 1362–77; Ejaz, N., Hamada, M., Diedrichsen, J. "Hand use predicts the structure of representations in sensorimotor cortex." *Nat Neurosci* 18 (2015): 1034–40.
23. Liepniann, H. "Apraxia." *Ergebnisse der Medizinischen Gesellschaft* I (1920): 516–43.
24. Elbert, T., Pantev, C., Wienbruch, C., et al. "Increased cortical representation of the fingers of the left hand in string players." *Science* 270 (1995): 305–7.
25. Blake, D. T., Byl, N. N., Cheung, S., et al. "Sensory representation abnormalities that parallel focal hand dystonia in a primate model." *Somatosens Mot Res* 19 (2002): 347–57.
26. Graziano, M. S., Cooke, D. F., Taylor, C. S. "Coding the location of the arm by sight." *Science* 290 (2000): 1782–86.
27. Hetu, S., Gregoire, M., Saimpont, A., et al. "The neural network of motor imagery: An ALE meta-analysis." *Neurosci Biobehav Rev* 37 (2013): 930–49.
28. Viswanathan, S., Fritz, C., Grafton, S. T. "Telling the right hand from the left hand. Multisensory integration, not motor imagery, solves the problem." *Psychol Sci* 23 (2012): 598–607.
29. Bassetti, C., Bogousslavsky, J., Regli, F. "Sensory syndromes in parietal stroke." *Neurology* 43 (1993): 1942–49.
30. Padberg, J., Recanzone, G., Engle, J., et al. "Lesions in posterior parietal

area 5 in monkeys result in rapid behavioral and cortical plasticity." *J Neurosci* 30 (2010): 12918–35.

31. Auerbach, B. M., Ruff, C. B. "Limb bone bilateral asymmetry: Variability and commonality among modern humans." *J Hum Evol* 50 (2006): 203–18.
32. Linkenauger, S. A., Witt, J. K., Bakdash, J. Z., et al. "Asymmetrical body perception: A possible role for neural body representations." *Psychol Sci* 20 (2009): 1373–80.
33. Fong, S. S., Chung, J. W., Ng, S. S., et al. "Differential postural control and sensory organization in young tennis players and taekwondo practitioners." *Motor Control* 18 (2014): 103–11.
34. Gerbino, P. G., Griffin, E. D., Zurakowski, D. "Comparison of standing balance between female collegiate dancers and soccer players." *Gait Posture* 26 (2007): 501–7.
35. Mouchnino, L., Aurenty, R., Massion, J., Pedotti, A. "Coordination between equilibrium and head-trunk orientation during leg movement: A new strategy build up by training." *J Neurophysiol* 67 (1992): 1587–98.
36. Bringoux, L., Marin, L., Nougier, V., et al. "Effects of gymnastics expertise on the perception of body orientation in the pitch dimension." *J Vestib Res* 10 (2000): 251–58.
37. Pick, A. *Über Störungen der Orientierung am eigenen Körper, Arbeiten aus der deutschen psychiatrischen Universitäts-klinik in Prag.* Berlin: Karger, 1908.
38. Sirigu, A., Grafman, J., Bressler, K., Sunderland, T. "Multiple representations contribute to body knowledge processing. Evidence from a case of autotopagnosia." *Brain* 114 (1991): 629–42.
39. Schwoebel, J., Coslett, H. B. "Evidence for multiple, distinct representations of the human body." *J Cogn Neurosci* 17 (2005): 543–53.
40. Podoll, K., Robinson, D. *Migraine Art: The Migraine Experience from Within.* Berkeley, CA: North Atlantic Books, 2009.
41. Woods, R. P., Iacoboni, M., Mazziotta, J. C. "Brief report: Bilateral spreading cerebral hypoperfusion during spontaneous migraine headache." *N Engl J Med* 331 (1994): 1689–92.
42. Leker, R. R., Karni, A., River, Y. "Microsomatoagnosia: Whole body schema illusion as part of an epileptic aura." *Acta Neurol Scand* 94 (1996): 383–85.
43. Hartmann, A. S., Thomas, J. J., Wilson, A. C., Wilhelm, S. "Insight impairment in body image disorders: Delusionality and overvalued ideas in anorexia nervosa versus body dysmorphic disorder." *Psychiatry Res* 210 (2013): 1129–35.
44. Keizer, A., Smeets, M. A., Dijkerman, H. C., et al. "Aberrant somatosensory perception in Anorexia Nervosa." *Psychiatry Res* 200 (2012): 530–37.

45. Keizer, A., Smeets, M. A., Dijkerman, H. C., et al. "Tactile body image disturbance in anorexia nervosa." *Psychiatry Res* 190 (2011): 115–20.
46. Keizer, A., Smeets, M. A., Dijkerman, H. C., et al. "Too fat to fit through the door: First evidence for disturbed body-scaled action in anorexia nervosa during locomotion." *PLoS One* 8 (2013): e64602.
47. van Dijk, M. T., van Wingen, G. A., van Lammeren, A., et al. "Neural basis of limb ownership in individuals with body integrity identity disorder." *PLoS One* 8 (2013): e72212.
48. Gay, J., Monsma, E. V., Torres-McGehee, T. "Give or take a few? Comparing measured and self-reported height and weight as correlates of social physique anxiety." *Res Q Exerc Sport* 80 (2009): 656–62; Jonnalagadda, S. S., Rosenbloom, C. A., Skinner, R. "Dietary practices, attitudes, and physiological status of collegiate freshman football players." *J Strength Cond Res* 15 (2001): 507–13; Torres-McGehee, T. M., Monsma, E. V., Dompier, T. P., Washburn, S. A. "Eating disorder risk and the role of clothing in collegiate cheerleaders' body images." *J Athl Train* 47 (2012): 541–48; Torres-McGehee, T. M., Monsma, E. V., Gay, J. L., et al. "Prevalence of eating disorder risk and body image distortion among National Collegiate Athletic Association Division I varsity equestrian athletes." *J Athl Train* 46 (2011): 431–37.

4. The Hidden Hand

1. In reference to the SpaceX Falcon Heavy rocket launch, as told to Dave Mosher, Elon Musk says SpaceX has "done everything you can think of" to prepare Falcon Heavy for launch. *Business Insider*, February 6, 2018.
2. Critchley, M. "Disorders of the body-image." In *The Parietal Lobe*. London: Edward Arnold, 1953.
3. Halligan, P. W., Marshall, J. C. "Supernumerary phantom limb after right hemispheric stroke." *J Neurol Neurosurg Psychiatry* 59 (1995): 341–42; Halligan, P. W., Marshall, J. C., Wade, D. T. "Three arms: A case study of supernumerary phantom limb after right hemisphere stroke." *J Neurol Neurosurg Psychiatry* 56 (1993): 159–66; McGonigle, D. J., Hänninen, R., Salenius, S., et al. "Whose arm is it anyway? An fMRI case study of supernumerary phantom limb." *Brain* 125 (2002): 1265–74.
4. Sperry, R. W. "Neural basis of the spontaneous optokinetic response produced by visual inversion." *J Comp Physiol Psychol* 43 (1950): 482–89.
5. Von Holst, E. "Relations between the central nervous system and the peripheral organs." *British J Animal Behav* 2 (1954): 86–89.
6. Helmholtz, H. *Handbuch der physiologischen Optik*. Leipzig: Voss, 1866.
7. Goodale, M. A., Pélisson, D., Prablanc, C. "Large adjustments in visually

guided reaching do not depend on vision of the hand or perception of target displacement." *Nature* 320 (1986): 748–50.

8. Blakemore, S. J., Wolpert, D., Frith, C. "Why can't you tickle yourself?" *Neuroreport* 11 (2000): R11–16; Blakemore, S. J., Wolpert, D. M., Frith, C. D. "Central cancellation of self-produced tickle sensation." *Nat Neurosci* 1 (1998): 635–40.
9. Baddeley, R. J., Ingram, H. A., Miall, R. C. "System identification applied to a visuomotor task: Near-optimal human performance in a noisy changing task." *J Neurosci* 23 (2003): 3066–75.
10. Association of Tennis Professionals Staff, "Aussie Groth Hits Speedy Serves." *ATP World,* May 5, 2012.
11. Desmurget, M., Grafton, S. "Forward modeling allows feedback control for fast reaching movements." *Trends Cogn Sci* 4 (2000): 423–31.
12. van Sonderen, J. F., Gielen, C. C., Denier van der Gon, J. J. "Motor programmes for goal-directed movements are continuously adjusted according to changes in target location." *Exp Brain Res* 78 (1989): 139–46.
13. Prablanc, C., Martin, O. "Automatic control during hand reaching at undetected two-dimensional target displacements." *J Neurophysiol* 67 (1992): 455–69; Prablanc, C., Pélisson, D., Goodale, M. A. "Visual control of reaching movements without vision of the limb. I. Role of extraretinal feedback of target position in guiding the hand." *Exp Brain Res* 62 (1986): 293–302.
14. Desmurget, M., Gaveau, V., Vindras, P., et al. "On-line motor control in patients with Parkinson's disease." *Brain* 127 (2004a): 1755–73; Keele, S. W., Posner, M. I. "Processing of visual feedback in rapid movements." *J Exp Psychol* 77 (1968): 155–58.
15. Desmurget, M., Grafton, S. T. "Feedback or feedforward control: End of a dichotomy." In *Taking Action: Cognitive Neuroscience Perspectives on Intentional Acts*. Ed. S. H. Johnson-Frey. Cambridge, MA: Bradford Press, 2003.
16. Desmurget, M., Epstein, C., Turner, R., et al. "Role of the posterior parietal cortex in updating reaching movements to a visual target." *Nat Neurosci* 2 (1999): 563–67.
17. London, B. M., Miller, L. E. "Responses of somatosensory area 2 neurons to actively and passively generated limb movements." *J Neurophysiol* 109 (2013): 1505–13.
18. Wolpert, D. M., Goodbody, S. J., Husain, M. "Maintaining internal representations: The role of the human superior parietal lobe." *Nat Neurosci* 1 (1998): 529–33.
19. Mulliken, G. H., Musallam, S., Andersen, R. A. "Forward estimation of movement state in posterior parietal cortex." *Proc Natl Acad Sci USA* 105 (2008): 8170–77.

5. Pulling Strings

1. Collodi, C. *The Adventures of Pinocchio.* Trans. Carol Della Chiesa. London: Macmillan, 1929.
2. Hochner, B. "An embodied view of octopus neurobiology." *Curr Biol* 22 (2012): R887–92.
3. Ibid.
4. Yekutieli, Y., Sagiv-Zohar, R., Aharonov, R., et al. "Dynamic model of the octopus arm. I. Biomechanics of the octopus reaching movement." *J Neurophysiol* 94 (2005): 1443–58.
5. Alstermark, B., Isa, T. "Circuits for skilled reaching and grasping." *Annu Rev Neurosci* 35 (2012): 559–78.
6. Abelew, T. A., Miller, M. D., Cope, T. C., Nichols, T. R. "Local loss of proprioception results in disruption of interjoint coordination during locomotion in the cat." *J Neurophysiol* 84 (2000): 2709–14.
7. van der Meer, A. L. H. "Keeping the arm in the limelight: Advanced visual control of arm movements in neonates." *Eur J Paediatr Neurol* 4 (1997b): 103–8.
8. Lew, A. R., Butterworth, G. "The effects of hunger on hand-mouth coordination in newborn infants." *Dev Psychol* 31 (1995): 456–63.
9. Nugent, M. M., Milner, T. E. "Segmental specificity in belly dance mimics primal trunk locomotor patterns." *J Neurophysiol* 117 (2017): 1100–111.
10. Gesslbauer, B., Hruby, L. A., Roche, A. D., et al. "Axonal components of nerves innervating the human arm." *Ann Neurol* 82 (2017): 396–408.
11. Hudson, H. M., Park, M. C., Belhaj-Saïf, A., Cheney, P. D. "Representation of individual forelimb muscles in primary motor cortex." *J Neurophysiol* 118 (2017): 47–63.
12. Georgopoulos, A. P., Schwartz, A. B., Kettner, R. E. "Neuronal population coding of movement direction." *Science* 233 (1986): 1416–19.
13. Griffin, D. M., Hoffman, D. S., Strick, P. L. "Corticomotoneuronal cells are 'functionally tuned.'" *Science* 350 (2015): 667–70.
14. Bernstein, N. A. *The Coordination and Regulation of Movements.* New York: Pergamon Press, 1967.
15. Feldman, A. G. "Once more on the equilibrium-point hypothesis (lambda model) for motor control." *J Mot Behav* 18 (1986): 17–54.
16. Kamakura, N., Matsuo, M., Ishii, H., et al. "Patterns of static prehension in normal hands." *Am J Occup Ther* 34 (1980): 437–45.
17. Santello, M., Flanders, M., Soechting, J. F. "Postural hand synergies for tool use." *J Neurosci* 18 (1998): 10105–15.
18. Macfarlane, N. B. W., Graziano, M. S. A. "Diversity of grip in *Macaca mulatta.*" *Exp Brain Res* 197 (2009): 255–68.

19. d'Avella, A., Saltiel, P., Bizzi, E. "Combinations of muscle synergies in the construction of a natural motor behavior." *Nat Neurosci* 6 (2003): 300–308.
20. Overduin, S. A., d'Avella, A., Roh, J., Bizzi, E. "Modulation of muscle synergy recruitment in primate grasping." *J Neurosci* 28 (2008): 880–92.
21. Leo, A., Handjaras, G., Bianchi, M., et al. "A synergy-based hand control is encoded in human motor cortical areas." *Elife* 5 (2016): e13420.
22. Sawers, A., Allen, J. L., Ting, L. H. "Long-term training modifies the modular structure and organization of walking balance control." *J Neurophysiol* 114 (2015): 3359–73.
23. Allen, J. L., McKay, J. L., Sawers, A., et al. "Increased neuromuscular consistency in gait and balance after partnered, dance-based rehabilitation in Parkinson's disease." *J Neurophysiol* 118 (2017): 363–73.
24. Verrel, J., Lövdén, M., Lindenberger, U. "Normal aging reduces motor synergies in manual pointing." *Neurobiol Aging* 33 (2012): 200.el–10.
25. Anan, M., Hattori, H., Tanimoto, K., et al. "The coordination of joint movements during sit-to-stand motion in old adults: The uncontrolled manifold analysis." *Phys Ther Res* 20 (2017): 44–50.
26. Sawers, A., Pai, Y. C., Bhatt, T., Ting, L. H. "Neuromuscular responses differ between slip-induced falls and recoveries in older adults." *J Neurophysiol* 117 (2017): 509–22.
27. Cheung, V. C., Piron, L., Agostini, M., et al. "Stability of muscle synergies for voluntary actions after cortical stroke in humans." *Proc Natl Acad Sci USA* 106 (2009): 19563–68.
28. Penfield, W., Boldrey, E. "Somatic motor and sensory representation in the cerebral cortex of man as studied by electrical stimulation." *Brain* 15 (1937): 389–443.
29. Grafton, S. T., Woods, R. P., Mazziotta, J. C., Phelps, M. E. "Somatotopic mapping of the primary motor cortex in humans: Activation studies with cerebral blood flow and positron emission tomography." *J Neurophysiol* 66 (1991): 735–43.
30. Grafton, S. T., Woods, R. P., Mazziotta, J. C. "Within-arm somatotopy in human motor areas determined by positron emission tomography imaging of cerebral blood flow." *Exp Brain Res* 95 (1993): 172–76.
31. Vanhaereny, M., d'Errico, F., Stringer, C., et al. "Middle Paleolithic shell beads in Israel and Algeria." *Science* 312 (2006): 1785–88.
32. Farrell, D. F., Burbank, N., Lettich, E., Ojemann, G. A. "Individual variation in human motor-sensory (rolandic) cortex." *J Clin Neurophysiol* 24 (2007): 286–93.
33. Hudson, Park, Belhaj-Saïf, Cheney. "Representation of individual forelimb muscles in primary motor cortex."

34. Sanes, J. N., Donoghue, J. P., Thangaraj, V., et al. "Shared neural substrates controlling hand movements in human motor cortex." *Science* 268 (1995): 1775–77.
35. Yokoi, A., Arbuckle, S. A., Diedrichsen, J. "The role of human primary motor cortex in the production of skilled finger sequences." *J Neurosci* 38 (2018): 1430–42.
36. Dum, R. P., Strick, P. L. "Frontal lobe inputs to the digit representations of the motor areas on the lateral surface of the hemisphere." *J Neurosci* 25 (2005): 1375–86.
37. Collinger, J. L., Wodlinger, B., Downey, J. E., et al. "High-performance neuroprosthetic control by an individual with tetraplegia." *Lancet* 381 (2013): 557–64; Hochberg, L. R., Bacher, D., Jarosiewicz, B., et al. "Reach and grasp by people with tetraplegia using a neurally controlled robotic arm." *Nature* 485 (2012): 372–75.
38. Schwartz, A. B. "Cortical neural prosthetics." *Annu Rev Neurosci* 27 (2004): 487–507.
39. Serruya, M. D., Hatsopoulos, N. G., Paninski, L., et al. "Instant neural control of a movement signal." *Nature* 416 (2002): 141–42.
40. Wymbs, N. F., Grafton, S. T. "The human motor system supports sequence-specific representations over multiple training-dependent timescales." *Cereb Cortex* 25 (2015): 4213–25.

6. Perspectives

1. Wrangell, Ferdinand von. *Narrative of an Expedition to the Polar Sea, in the years 1820, 1821, 1822 and 1823.* New York: Harper & Brothers, 1855, 289.
2. Doyle, J. C., Francis, B. A., Tannenbaum, A. R. *Feedback Control Theory.* Mineola, NY: Dover Books on Electrical Engineering, 2009.
3. Souman, J. L., Frissen, I., Sreenivasa, M. N., Ernst, M. O. "Walking straight into circles." *Curr Biol* 19 (2009): 1538–42.
4. Cheung, A., Vickerstaff, R. "Finding the way with a noisy brain." *PLoS Comput Biol* 6 (2010): e1000992.
5. Ibid.
6. Darwin, C. "Origin of certain instincts." *Nature* 7 (1873): 417–18.
7. Wolbers, T., Hegarty, M. "What determines our navigational abilities?" *Trends Cogn Sci* 14 (2010): 138–46.
8. O'Keefe, J., Burgess, N. "Geometric determinants of the place fields of hippocampal neurons." *Nature* 381 (1996): 425–28.
9. Zhang, H., Zherdeva, K., Ekstrom, A. D. "Different 'routes' to a cognitive map: Dissociable forms of spatial knowledge derived from route and cartographic map learning." *Mem Cognit* 42 (2014): 1106–17.

10. He, Q., McNamara, T. P. "Spatial updating strategy affects the reference frame in path integration." *Psychon Bull Rev* in press (2017): 1–7.
11. Harris, M. A., Wolbers, T. "How age-related strategy switching deficits affect wayfinding in complex environments." *Neurobiol Aging* 35 (2014): 1095–102.
12. Schuck, N. W., Doeller, C. F., Polk, T. A., et al. "Human aging alters the neural computation and representation of space." *Neuroimage* 117 (2015): 141–50.
13. Wiener, J. M., de Condappa, O., Harris, M. A., Wolbers, T. "Maladaptive bias for extrahippocampal navigation strategies in aging humans." *J Neurosci* 33 (2013): 6012–17.
14. Beurze, S. M., Toni, I., Pisella, L., Medendorp, W. P. "Reference frames for reach planning in human parietofrontal cortex." *J Neurophysiol* 104 (2010): 1736–45.
15. Giller, C. A., Mornet, P., Moreau, J.-F. "The first formulation of image-based stereotactic principles: The forgotten work of Gaston Contremoulins." *J Neurosurg* (2017): 1–10.
16. Batista, A. P., Santhanam, G., Yu, B. M., et al. "Reference frames for reach planning in macaque dorsal premotor cortex." *J Neurophysiol* 98 (2007): 966–83.
17. Bernier, P.-M., Grafton, S. T. "Human posterior parietal cortex flexibly determines reference frames for reaching based on sensory context." *Neuron* 68 (2010): 776–88.
18. Niehof, N., Tramper, J. J., Doeller, C. F., Medendorp, W. P. "Updating of visual orientation in a gravity-based reference frame." *J Vis* 17 (2017): 1–10.
19. Belmonti, V., Cioni, G., Berthoz, A. "Anticipatory control and spatial cognition in locomotion and navigation through typical development and in cerebral palsy." *Dev Med Child Neurol* 58 suppl. 4 (2016): 22–27.
20. Perennou, D., Piscicelli, C., Barbieri, G., et al. "Measuring verticality perception after stroke: Why and how?" *Neurophysiol Clin* 44 (2014): 25–32.
21. Burns, J. K., Blohm, G. "Multi-sensory weights depend on contextual noise in reference frame transformations." *Front Hum Neurosci* 4 (2010): 221; Tagliabue, M., Arnoux, L., McIntyre, J. "Keep your head on straight: Facilitating sensori-motor transformations for eye-hand coordination." *Neuroscience* 248 (2013): 88–94.
22. Barra, J., Marquer, A., Joassin, R., et al. "Humans use internal models to construct and update a sense of verticality." *Brain* 133 (2010): 3552–63.
23. Teixeira, L. A., de Oliveira, D. L., Romano, R. G., Correa, S. C. "Leg preference and interlateral asymmetry of balance stability in soccer players." *Res Q Exerc Sport* 82 (2011): 21–27.

24. Paillard, T., Bizid, R., Dupui, P. "Do sensorial manipulations affect subjects differently depending on their postural abilities?" *Br J Sports Med* 41 (2007): 435–38.
25. Pouget, A., Sejnowski, T. J. "Spatial transformations in the parietal cortex using basis functions." *J Cogn Neurosci* 9 (1997): 222–37.
26. Clarke, A. M., Öğmen, H., Herzog, M. H. "A computational model for reference-frame synthesis with applications to motion perception." *Vision Res* 126 (2016): 242–53.
27. Tanaka, H., Sejnowski, T. J. "Motor adaptation and generalization of reaching movements using motor primitives based on spatial coordinates." *J Neurophysiol* 113 (2015): 1217–33.
28. Tramper, J. J., Medendorp, W. P. "Parallel updating and weighting of multiple spatial maps for visual stability during whole body motion." *J Neurophysiol* 114 (2015): 3211–19.
29. Vickers, J. N. "Gaze of Olympic speedskaters skating at full speed on a regulation oval: Perception-action coupling in a dynamic performance environment." *Cogn Process* 7 (2006): 102–5.
30. Avillac, M., Denève, S., Olivier, E., et al. "Reference frames for representing visual and tactile locations in parietal cortex." *Nat Neurosci* 8 (2005): 941–49.
31. McGuire, L. M., Sabes, P. N. "Sensory transformations and the use of multiple reference frames for reach planning." *Nat Neurosci* 12 (2009): 1056–61.
32. Tseng, C.-H., Chow, H. M., Spillmann, L. "Falling skyscrapers: When cross-modal perception of verticality fails." *Psychol Sci* 24 (2013): 1341–47.

7. Learning to Solve Problems

1. Foderaro, L. W. "Bear-proof can is pop-top picnic for a crafty thief." *The New York Times,* July 24, 2009.
2. Herrmann, E., Hare, B., Cissewski, J., Tomasello, M. "A comparison of great apes' and human children's temperament." *Dev Sci* 14 (2011): 1393–405.
3. Bruner, J. "Organization of early skilled action." *Child Dev* 44 (1973): 1–11.
4. Rovee, C. K., Rovee, D. T. "Conjugate reinforcement of infant exploratory behavior." *J Exp Child Psychol* 8 (1969): 33–39.
5. Sutton, R. S., Barto, A. G. *Reinforcement Learning.* Cambridge, MA: MIT Press, 1998.
6. Engel, K. C., Flanders, M., Soechting, J. F. "Anticipatory and sequential motor control in piano playing." *Exp Brain Res* 113 (1997): 189–99.

7. Keen, R. "The development of problem solving in young children: A critical cognitive skill." *Annu Rev Psychol* 62 (2011): 1–21.
8. Keen, R., Lee, M.-H., Adolph, K. "Planning an action: A developmental progression in tool use." *Ecol Psychol* 26 (2014): 98–108.
9. Köhler, W. *The Mentality of Apes*. New York: Harcourt, 1927.
10. Cushman, F., Morris, A. "Habitual control of goal selection in humans." *Proc Natl Acad Sci USA* 112 (2015): 13817–22.
11. Cooper, R., Shallice, T. "Hierarchical schemas and goals in the control of sequential behavior." *Psychol Rev* 113 (2006): 887–916.
12. Rose, N. S., Luo, L., Bialystok, E., et al. "Cognitive processes in the Breakfast Task: Planning and monitoring." *Can J Exp Psychol* 69 (2015): 252–63.
13. Balaguer, J., Spiers, H., Hassabis, D., Summerfield, C. "Neural mechanisms of hierarchical planning in a virtual subway network." *Neuron* 90 (2016): 893–903.
14. Chen, X., Hertzog, C., Park, D. C. "Cognitive predictors of everyday problem solving across the lifespan." *Gerontology* 63 (2017): 372–74.
15. Shallice, T., Burgess, P. W. "Deficits in strategy application following frontal lobe damage in man." *Brain* 114 (1991): 727–41.
16. Clerget, E., Andres, M., Olivier, E. "Deficit in complex sequence processing after a virtual lesion of left BA45." *PLoS One* 8 (2013): e63722.

8. Purpose

1. Johns, J. "Sketchbook Notes." Art and Literature (Lausanne) 4 (Spring 1965): 192.
2. Darwin, C. *The Formation of Vegetable Mould Through the Action of Worms, with Observations on Their Habits*. London: John Murray, 1881.
3. Crickette, M. S., Call, J., Boesch, C. *Tool Use in Animals: Cognition and Ecology*. Cambridge: Cambridge University Press, 2013; Shumaker, R. W., Walkup, K. R., Beck, B. B. *Animal Tool Behavior: The Use and Manufacture of Tools by Animals*. Baltimore: Johns Hopkins University Press, 2011.
4. Simler, J. *Die Alpen De Alpibus Commentarius*. München: Gesellschaft Alpiner Bücherfreunde, 1931.
5. Cardinali, L., Frassinetti, F., Brozzoli, C., et al. "Tool-use induces morphological updating of the body schema." *Curr Biol* 19 (2009): R478–79.
6. Iriki, A., Tanaka, M., Iwamura, Y. "Coding of modified body schema during tool use by macaque postcentral neurones." *Neuroreport* 7 (1996): 2325–30.
7. Arbib, M. A., Bonaiuto, J. B., Jacobs, S., Frey, S. H. "Tool use and the distalization of the end-effector." *Psychol Res* 73 (2009): 441–62.

8. Bonifazi, S., Farne, A., Rinaldesi, L., Ladavas, E. "Dynamic size-change of peri-hand space through tool-use: Spatial extension or shift of the multi-sensory area." *J Neuropsychol* 1 (2007): 101–14.
9. Thieme, H. "Lower Palaeolithic hunting spears from Germany." *Nature* 385 (1997): 807–10.
10. Povinelli, D. J., Reaux, J. E., Frey, S. H. "Chimpanzees' context-dependent tool use provides evidence for separable representations of hand and tool even during active use within peripersonal space." *Neuropsychologia* 48 (2010): 243–47.
11. Bril, B., Foucart, J. *Enacting the Perception of the Affordances of Potential Tools. I: The Case of Children Hammering.* In *Studies in Perception and Action VIII,* ed. H. Heft, K. L. Marsh. Mahwah, NJ: Lawrence Erlbaum Associates, Inc., 2005.
12. Bril, B., Dietrich, G., Foucart, J., et al. "Tool use as a way to assess cognition: How do captive chimpanzees handle the weight of the hammer when cracking a nut?" *Anim Cogn* 12 (2009): 217–35.
13. Mangalam, M., Fragaszy, D. M. "Wild bearded capuchin monkeys crack nuts dexterously." *Curr Biol* 25 (2015): 1334–39.
14. Sakata, H., Taira, M., Murata, A., Mine, S. "Neural mechanisms of visual guidance of hand action in the parietal cortex of the monkey." *Cereb Cortex* 5 (1995): 429–38.
15. Rizzolatti, G., Gentilucci, M., Camarda, R. M., et al. "Neurons related to reaching-grasping arm movements in the rostral part of area 6 (area 6a beta)." *Exp Brain Res* 82 (1990): 337–50.
16. Matelli, M., Camarda, M., Glickstein, M., Rizzolatti, G. "Afferent and efferent projections of the inferior area 6 in the macaque monkey." *J Comp Neurol* 251 (1986): 281–98.
17. Frey, S. H., Vinton, D., Norlund, R., Grafton, S. T. "Cortical topography of human anterior intraparietal cortex active during visually-guided grasping." *Cogn Brain Res* 23 (2005): 397–405; Grafton, S. T., Arbib, M. A., Fadiga, L., Rizzolatti, G. "Localization of grasp representations in humans by positron emission tomography. 2. Observation compared with imagination." *Exp Brain Res* 112 (1996): 103–11; Grafton, S. T., Mazziotta, J. C., Woods, R. P., Phelps, M. E. "Human functional anatomy of visually guided finger movements." *Brain* 115 (1992): 565–87.
18. Jeannerod, M., Arbib, M. A., Rizzolatti, G., Sakata, H. "Grasping objects: the cortical mechanisms of visuomotor transformation." *Trends Neurosci* 18 (1995): 314–20.
19. Sangole, A. P., Levin, M. F. "Palmar arch dynamics during reach-to-grasp tasks." *Exp Brain Res* 190 (2008): 443–52.
20. Ambrosini, E., Reddy, V., de Looper, A., et al. "Looking ahead:

Anticipatory gaze and motor ability in infancy." *PLoS One* 8 (2013): e67916.

21. Grafton, S. T., Fadiga, L., Arbib, M. A., Rizzolatti, G. "Premotor cortex activation during observation and naming of familiar tools." *Neuroimage* 6 (1997): 231–36.
22. Costantini, M., Ambrosini, E., Scorolli, C., Borghi, A. M. "When objects are close to me: Affordances in the peripersonal space." *Psychon Bull Rev* 18 (2011): 302–8; Handy, T. C., Grafton, S. T., Shroff, N., et al. "Graspable objects grab attention when the potential for action is recognized." *Nat Neurosci* 6 (2003): 421–27.
23. Proverbio, A. M., Adorni, R., D'Aniello, G. E. "250 ms to code for action affordance during observation of manipulable objects." *Neuropsychologia* 49 (2011): 2711–17.
24. Marneweck, M., Knelange, E., Lee-Miller, T., et al. "Generalization of dexterous manipulation is sensitive to the frame of reference in which it is learned." *PLoS One* 10 (2015): e0138258.
25. Sakaguchi, Y., Ishida, F., Shimizu, T., Murata, A. "Time course of information representation of macaque AIP neurons in hand manipulation task revealed by information analysis." *J Neurophysiol* 104 (2010): 3625–43.
26. Povinelli, D. J. *World Without Weight: Perspectives on an Alien Mind.* Oxford: Oxford University Press, 2012.
27. Harmand, S., Lewis, J. E., Feibel, C. S., et al. "3.3-million-year-old stone tools from Lomekwi 3, West Turkana, Kenya." *Nature* 521 (2015): 310–15.
28. Gallivan, J. P., Cant, J. S., Goodale, M. A., Flanagan, J. R. "Representation of object weight in human ventral visual cortex." *Curr Biol* 24 (2014): 1866–73.
29. Marzke, M. W. "Precision grips, hand morphology, and tools." *Am J Phys Anthropol* 102 (1997): 91–110.
30. Almécija, S., Moyà-Solà, S., Alba, D. M. "Early origin for human-like precision grasping: A comparative study of pollical distal phalanges in fossil hominins." *PLoS One* 5 (2010): e11727.
31. Kivell, T. L., Kibii, J. M., Churchill, S. E., et al. "*Australopithecus sediba* hand demonstrates mosaic evolution of locomotor and manipulative abilities." *Science* 333 (2011): 1411–17.
32. Liepmann, H. "The left hemisphere and action." In *Translations from Liepmann's Essays on Apraxia.* Research Bulletin 506. London, ONT: Department of Psychology, University of Western Ontario, 1980. Translated from *Munch Med Wschr* (1905): 48–49.
33. Haaland, K. Y., Harrington, D. L., Knight, R. T. "Neural representations of skilled movement." *Brain* 123, pt. 11 (2000): 2306–13.
34. Johnson-Frey, S. H., Newman-Norlund, R., Grafton, S. T. "A distributed

left hemisphere network active during planning of everyday tool use skills." *Cereb Cortex* 15 (2005): 681–95.

9. Costs

1. Xenophon. *Anabasis.* Book VI, section IV. Trans. H. G. Dakyns. London Macmillan and Company, 1901.
2. Pontzer, H. "Economy and endurance in human evolution." *Curr Biol* 27 (2017): 613–21.
3. Saint-Exupéry, A. de, Galantiere, L. T. *Wind, Sand and Stars.* San Diego: Harcourt, 2002.
4. Rubenson, J., Heliams, D. B., Maloney, S. K., et al. "Reappraisal of the comparative cost of human locomotion using gait-specific allometric analyses." *J Exp Biol* 210 (2007): 3513–24.
5. Pontzer, H., Raichlen, D. A., Wood, B. M., et al. "Hunter-gatherer energetics and human obesity." *PLoS One* 7 (2012): e40503.
6. Ivanenko, Y. P., Dominici, N., Lacquaniti, F. "Development of independent walking in toddlers." *Exerc Sport Sci Rev* 35 (2007): 67–73.
7. Zarrugh, M. Y., Radcliffe, C. W. "Predicting metabolic cost of level walking." *Eur J Appl Physiol Occup Physiol* 38 (1978): 215–23.
8. Selinger, J. C., O'Connor, S. M., Wong, J. D., Donelan, J. M. "Humans can continuously optimize energetic cost during walking." *Curr Biol* 25 (2015): 2452–56.
9. Vernillo, G., Savoldelli, A., Skafidas, S., et al. "An extreme mountain ultra-marathon decreases the cost of uphill walking and running." *Front Physiol* 7 (2016): 530.
10. Brown, M., Hislop, H. J., Waters, R. L., Porell, D. "Walking efficiency before and after total hip replacement." *Phys Ther* 60 (1980): 1259–63.
11. Huang, T. W., Shorter, K. A., Adamczyk, P. G., Kuo, A. D. "Mechanical and energetic consequences of reduced ankle plantar-flexion in human walking." *J Exp Biol* 218 (2015): 3541–50.
12. Huang, H. J., Ahmed, A. A. "Reductions in muscle coactivation and metabolic cost during visuomotor adaptation." *J Neurophysiol* 112 (2014): 2264–74.
13. Holmes, G. "The symptoms of acute cerebellar injuries due to gunshot injuries." *Brain* 40 (1917): 461–535.
14. Thach, W. T. "Correlation of neural discharge with pattern and force of muscular activity, joint position, and direction of intended next movement in motor cortex and cerebellum." *J Neurophysiol* 41 (1978): 654–76.
15. Bernard, J. A., Seidler, R. D. "Moving forward: Age effects on the cerebellum underlie cognitive and motor declines." *Neurosci Biobehav Rev* 42 (2014): 193–207.

16. Marshall, J. M. "Peripheral chemoreceptors and cardiovascular regulation." *Physiol Rev* 74 (1994): 543–94.
17. Miura, M., Reis, D. J. "The role of the solitary and paramedian reticular nuclei in mediating cardiovascular reflex responses from carotid baro- and chemoreceptors." *J Physiol* 223 (1972): 525–48.
18. Gerrits, N. M., Voogd, J. "The nucleus reticularis tegmenti pontis and the adjacent rostral paramedian reticular formation: Differential projections to the cerebellum and the caudal brain stem." *Exp Brain Res* 62 (1986): 29–45.
19. Wong, J. D., O'Connor, S. M., Selinger, J. C., Donelan, J. M. "Contribution of blood oxygen and carbon dioxide sensing to the energetic optimization of human walking." *J Neurophysiol* 118 (2017): 1425–33.
20. Ito, M. "Neural design of the cerebellar motor control system." *Brain Res* 40 (1972): 81–102.
21. Todorov, E., Jordan, M. I. "Smoothness maximization along a predefined path accurately predicts the speed profiles of complex arm movements." *J Neurophysiol* 80 (1998): 696–714.
22. Harris, C. M., Wolpert, D. M. "Signal-dependent noise determines motor planning." *Nature* 394 (1998): 780–84.
23. Voloshina, A. S., Ferris, D. P. "Biomechanics and energetics of running on uneven terrain." *J Exp Biol* 218 (2015): 711–19.
24. Simpson, K. M., Munro, B. J., Steele, J. R. "Backpack load affects lower limb muscle activity patterns of female hikers during prolonged load carriage." *J Electromyogr Kinesiol* 21 (2011): 782–88.
25. Hayhoe, M., Gillam, B., Chajka, K., Vecellio, E. "The role of binocular vision in walking." *Vis Neurosci* 26 (2009): 73–80.
26. Turner, R. S., Desmurget, M. "Basal ganglia contributions to motor control: A vigorous tutor." *Curr Opin Neurobiol* 20 (2010): 704–16.
27. DeLong, M. R., Alexander, G. E., Georgopoulos, A. P., et al. "Role of basal ganglia in limb movements." *Human Neurobiol* 2 (1984): 235–44.
28. Desmurget, M., Grafton, S. T., Vindras, P., et al. "Basal ganglia network mediates the control of movement amplitude." *Exp Brain Res* 153 (2003): 197–209.
29. Desmurget, M., Grafton, S. T., Vindras, P., et al. "The basal ganglia network mediates the planning of movement amplitude." *Eur J Neurosci* 19 (2004b): 2871–80.
30. Turner, R. S., Desmurget, M., Grethe, J., et al. "Motor subcircuits mediating the control of movement extent and speed." *J Neurophysiol* 90 (2003): 3958–66; Turner, R. S., Grafton, S. T., Votaw, J. R., et al. "Motor

subcircuits mediating the control of movement velocity: A PET study." *J Neurophysiol* 80 (1998): 2162–76.

31. Vaillancourt, D. E., Mayka, M. A., Thulborn, K. R., Corcos, D. M. "Subthalamic nucleus and internal globus pallidus scale with the rate of change of force production in humans." *Neuroimage* 23 (2004): 175–86.
32. Turner, Desmurget. "Basal ganglia contributions to motor control: A vigorous tutor."
33. Mazzoni, P., Hristova, A., Krakauer, J. W. "Why don't we move faster? Parkinson's disease, movement vigor, and implicit motivation." *J Neurosci* 27 (2007): 7105–16.
34. Gepshtein, S., Li, X., Snider, J., et al. "Dopamine function and the efficiency of human movement." *J Cogn Neurosci* 26 (2014): 645–57.
35. Christie, C. J., Scott, P. A. "Metabolic responses of South African soldiers during simulated marching with 16 combinations of speed and backpack load." *Mil Med* 170 (2005): 619–22.

10. Of One Mind

1. Quoted in Pfitzinger, P., Douglas, S. *Advanced Marathoning.* Champaign, IL: Human Kinetics, 2008. Known as the Flying Finn, Nurmi was the winner of nine gold medals in distance running.
2. Mosso, A. *Fatigue.* London: Allen & Unwin Ltd., 1915.
3. Hill, A. V. The Mechanisms of Muscular Contraction. The Nobel Prize Lecture 1923 (Les Prix Nobel en 1923), Imprimerie Royale. P. A. Norstedt & Fils, 1924.
4. Bainbridge, F. A. *The Physiology of Muscular Exercise.* London: Longmans, Green and Co., 1931.
5. Noakes, T. D. "Fatigue is a brain-derived emotion that regulates the exercise behavior to ensure the protection of whole body homeostasis." *Front Physiol* 3 (2012): 82.
6. Mosso. *Fatigue,* 156.
7. Bainbridge. *The Physiology of Muscular Exercise.*
8. Barwood, M. J., Weston, N. J., Thelwell, R., Page, J. "A motivational music and video intervention improves high-intensity exercise performance." *J Sports Sci Med* 8 (2009): 435–42.
9. Morton, R. H. "Deception by manipulating the clock calibration influences cycle ergometer endurance time in males." *J Sci Med Sport* 12 (2009): 332–37.
10. Swart, J., Lamberts, R. P., Lambert, M. I., et al. "Exercising with reserve: Evidence that the central nervous system regulates prolonged exercise performance." *Br J Sports Med* 43 (2009a): 782–88.

11. Swart, J., Lamberts, R. P., Lambert, M. I., et al. "Exercising with reserve: Exercise regulation by perceived exertion in relation to duration of exercise and knowledge of endpoint." *Br J Sports Med* 43 (2009b): 775–81.
12. Parvizi, J., Rangarajan, V., Shirer, W. R., et al. "The will to persevere induced by electrical stimulation of the human cingulate gyrus." *Neuron* 80 (2013): 1359–67.
13. Markov, N. T., Ercsey-Ravasz, M. M., Ribeiro Gomes, A. R., et al. "A weighted and directed interareal connectivity matrix for macaque cerebral cortex." *Cereb Cortex* 24 (2014): 17–36.
14. Park, S. Q., Kahnt, T., Rieskamp, J., Heekeren, H. R. "Neurobiology of value integration: When value impacts valuation." *J Neurosci* 31 (2011): 9307–14.
15. Volz, L. J., Welborn, B. L., Gobel, M. S., et al. "Harm to self outweighs benefit to others in moral decision making." *Proc Natl Acad Sci USA* 114 (2017): 7963–68.
16. Amemori, K.-I., Amemori, S., Graybiel, A. M. "Motivation and affective judgments differentially recruit neurons in the primate dorsolateral prefrontal and anterior cingulate cortex." *J Neurosci* 35 (2015): 1939–53.
17. Amemori, K., Graybiel, A. M. "Localized microstimulation of primate pregenual cingulate cortex induces negative decision-making." *Nat Neurosci* 15 (2012): 776–85.
18. Mayberg, H. S., Lozano, A. M., Voon, V., et al. "Deep brain stimulation for treatment-resistant depression." *Neuron* 45 (2005): 651–60.
19. St. Clair Gibson, A., De Koning, J. J., Thompson, K. G., et al. "Crawling to the finish line: Why do endurance runners collapse? Implications for understanding of mechanisms underlying pacing and fatigue." *Sports Med* (Auckland) 43 (2013): 413–24.
20. Cannon, W. B. "'Voodoo' Death." *Am Anthropol* 44 (1942): 169–81.
21. Samuels, M. A. "The brain-heart connection." *Circulation* 116 (2007): 77–84.
22. Krause, T., Werner, K., Fiebach, J. B., et al. "Stroke in right dorsal anterior insular cortex is related to myocardial injury." *Ann Neurol* 81 (2017): 502–11.
23. Critchley, H. D., Garfinkel, S. N. "Interactions between visceral afferent signaling and stimulus processing." *Front Neurosci* 9 (2015): 286.
24. Critchley, H. D., Mathias, C. J., Josephs, O., et al. "Human cingulate cortex and autonomic control: Converging neuroimaging and clinical evidence." *Brain* 126 (2003): 2139–52.
25. Lacuey, N., Hampson, J. P., Theeranaew, W., et al. "Cortical structures associated with human blood pressure control." *JAMA Neurol* 75 (2018): 194–202.

26. Dum, R. P., Levinthal, D. J., Strick, P. L. "Motor, cognitive, and affective areas of the cerebral cortex influence the adrenal medulla." *Proc Natl Acad Sci USA* 113 (2016): 9922–27.
27. Cieslak, M., Ryan, W. S., Babenko, V., et al. "Quantifying rapid changes in cardiovascular state with a moving ensemble average." *Psychophysiology* 55 (2018): e13018; Cieslak, M., Ryan, W. S., Macy, A., et al. "Simultaneous acquisition of functional magnetic resonance images and impedance cardiography." *Psychophysiology* 52 (2015): 481–88.
28. Stine-Morrow, E. A. L., Payne, B. R., Roberts, B. W., et al. "Training versus engagement as paths to cognitive enrichment with aging." *Psychol Aging* 29 (2014): 891–906.
29. Prakash, R. S., Voss, M. W., Erickson, K. I., Kramer, A. F. "Physical activity and cognitive vitality." *Annu Rev Psychol* 66 (2015): 769–97.

Index

ABOUT THE AUTHOR

Scott Grafton holds the Bedrosian Coyne Presidential Chair in Neuroscience at the University of California, Santa Barbara. He is director of the UCSB Brain Imaging Center and codirector of the Institute for Collaborative Biotechnologies, which draws on bio-inspiration and innovative bioengineering solutions for both nonmedical and medical challenges posed by the defense and medical communities. He received his medical degree from the University of Southern California and specialty training in neurology and nuclear medicine. He is an avid alpinist with many high-altitude ascents and wilderness meanderings, but he makes his home in Santa Barbara.

A NOTE ON THE TYPE

The text of this book was set in Plantin, a typeface first cut in 1913 by the Monotype Corporation of London. Though the face bears the name of the great Christopher Plantin (ca. 1520–1589), who in the latter part of the sixteenth century owned, in Antwerp, the largest printing and publishing firm in Europe, it is a rather free adaptation of designs by Claude Garamond made for that firm. With its strong, simple lines, Plantin is a no-nonsense face of exceptional legibility.

Composed by North Market Street Graphics
Lancaster, Pennsylvania

Printed and bound by Berryville Graphics
Berryville, Virginia

Designed by Michael Collica